The Fundamentals of Algebra Workbook

by

John Gorham

© Copyright 1996. John Gorham. All rights reserved.

ISBN 0-9637658-3-3

Third printing September 2003

No part of this publication may be reproduced, stored in a retrieval system, or transmitted in any form, or by any means: electronic, electrostatic, magnetic tapes, mechanical photocopying, recording, or otherwise, without prior and current written permission from the publisher.

Published by Davis-Gorham Press
6310 Ramsgate Court
Brentwood, Tennessee 37027

Contents

Chapter 1: Properties, Signs, & Order of Operation

Definitions 1
The Real Number System 2
Properties 3
Sign: Addition 8
         Subtraction 10
         Multiplication 11
         Division 12
Order of Operation 13
Chapter Review 21

Chapter 2: Addition and Subtraction of Polynomials

Addition 23
Subtraction 26
Chapter Review 29
Cumulative Review 30

Chapter 3: Multiplication of Polynomials

Monomials 31
Monomials and Polynomials 34
Binomials and Binomials 36
Binomials and Trinomials 39
Chapter Review 40
Cumulative Review 41

Chapter 4: Division of Polynomials & Negative Exponents

Zero Exponents 43
Monomials Divided into Monomials 43
Monomials Divided into Polynomials 46
Binomials Divided into Trinomials 47
Binomials Divided into Polynomials 51
Negative Exponents 54
Chapter Review 58
Cumulative Review 59

Chapter 5: Solving Equations

- Removing the Parentheses 61
- Combining Like Terms 62
- One Step Equations 63
- Two Step Equations 67
- Multi-Term Equations 70
- Absolute Value Equations 72
- Chapter Review 74
- Cumulative Review 75

Chapter 6: Inequalities

- Reading Inequalities 77
- Graphing Inequalities 78
- The Solution Set 79
- Solving Inequalities 82
- Compound Inequalities 84
- Chapter Review 86
- Cumulative Review 87

Chapter 7 & 8: Factoring

- Checking Factors 89
- Common Factors 90
- The Difference Between Two Perfect Squares 91
- Trinomials That Factors as a Binomial Square 93
- Regrouping 95
- Trinomials 99
- Combination Problems 110
- Chapter Review 113
- Cumulative Review 114

Chapter 9: Quadratic Equations and Rational Expressions

- Solving Quadratic Equations by Factoring 115
- Simplifying Rational Expressions 118
- Multiplication of Rational Expressions 121
- Division of Rational Expressions 124
- Chapter Review 127
- Cumulative Review 128

Chapter 10: Addition & Subtraction of Rational Expressions

- Fractions with the Same Denominator 129
- Finding the Least Common Denominator 133
- Addition and Subtraction of Rational Expressions 136
- Rational Equations 142
- Chapter Review 150
- Cumulative Review 152

Chapter 11: Solving Linear Equations

- Standard and Slope-Intercept Forms 153
- Finding Coordinates 155
- Graphing Coordinates 157
- Graphing a Linear Equation 158
- Solving Simultaneous Equations by Graphing 160

Solving Simultaneous Equations by Addition ... 163
Solving Simultaneous Equations by Substitution ... 169
Solving Simultaneous Inequalities by Graphing ... 173
Chapter Review ... 176
Cumulative Review ... 177

Chapter 12: Slope and Writing Equations

Describing the Slope ... 179
Finding the Slope: Four Methods ... 180
Writing Linear Equations ... 184
Equal, Parallel, and Perpendicular Linear Equations ... 187
Chapter Review ... 188
Cumulative Review ... 189

Chapter 13: Radicals

The Square Root ... 191
Simplifying Radicals ... 192
Addition of Radicals ... 193
Multiplication of Radicals ... 194
Division and Rationalizing Radicals ... 198
Simplifying and Combining Radicals ... 203
Radical Equations ... 205
Chapter Review ... 207
Cumulative Review ... 208

Chapter 14: Quadratic Equations

Solving Quadratic Equations by Factoring ... 209
The Quadratic Formula ... 210
Solving Quadratic Equations by Completing the Binomial Square ... 212
Writing Quadratic Equations ... 217
Chapter Review ... 218
Cumulative Review ... 220

Chapter 15: Math Riddles

Introduction ... 221
One Solution Math Riddles ... 223
Two Solution Math Riddles ... 236
Coin Riddles ... 248
Distance Riddles ... 250
Interest Riddles ... 252
Mixture Riddles ... 254
Chapter Review ... 256
Cumulative Review ... 258

Chapter 1
Properties, Signs, Order of Operation

Definitions

Algebra - The mathematical science that teaches the skills needed to solve equations.
Variable - A letter that can have different values. In $6a^2$, **a** is the variable.
Numerical Coefficient - A number in front of a variable. In $6a^2$, **6** is the numerical coefficient.
Exponent - A number to the right of and above a number or variable. In $6a^2$, the **2** is the exponent. The exponent is also called the power.
Term - A number, a variable, the product of quotient of numbers and/or variables. Examples: 3; y; 2x; $3y^4z$.
Algebraic Expression - A mathematical statement which contains a variable. Examples: 4x - 1; a + 5b + 8c
Equation - A mathematical statement containing an equal sign. Most equations are also algebraic expressions. Example: 2x - 3 = 7.
Absolute Value - The value of a number without any sign. The symbols for absolute value are two vertical bars. Example: |-2| = 2

Identify that which is underlined:

1. <u>4</u>x^2 _____
2. 5<u>y</u>^{3} _____
3. <u>x</u> = 8 _____
4. <u>6</u>x^6 _____
5. <u>y - 3z</u> _____
6. 3$c^{\underline{4}}$ _____
7. <u>2</u>y^5 _____
8. <u>4y + 7</u> _____
9. <u>7</u>xz _____
10. <u>a = c</u> _____

12. 4<u>x</u>^{2} _____
13. <u>7</u>ab _____
14. <u>| +4 |</u> _____
15. 3$a^{\underline{4}}$ _____
16. <u>5x = 40</u> _____
17. <u>3c - 2</u> _____
18. <u>22</u>a _____
19. <u>5</u>yz^2 _____
20. 6<u>a</u> _____
21. <u>|-7|</u> _____

23. 4$x^{\underline{2}}$ _____
24. <u>7y + 2 - x</u> _____
25. <u>4</u>x _____
26. <u>10y = 35</u> _____
27. <u>21</u>y _____
28. <u>5</u>b _____
29. <u>| +11 |</u> _____
30. <u>13</u>a^3 _____
31. <u>R = 0</u> _____
32. <u>16 - x^2</u> _____

The *Real Number* System

The **real numbers** are made up of five groups of numbers. They are:
1. **Natural or counting numbers** - The numbers you use when you count, such as 1, 2, 3, ect.
2. **Whole numbers** - This group adds the number zero, 0, to the group of counting numbers. Example: 0, 1, 2, 3.......
3. **Integers** - This group adds all the negative numbers to the group of whole numbers. Example: .... -2, -1, 0, 1, 2,.....
4. **Rational numbers** - This group includes any number that can be put in the form of a fraction. All counting numbers, whole numbers, and integers are also rational numbers. Examples are: 4, -3, $\frac{3}{8}$, .125, $3\frac{2}{3}$, .33̄3̄
5. **Irrational numbers** - Square roots ($\sqrt{\ }$) that are not perfect squares, and the non-repeating decimals are included in this group. Some examples are: .792..., $\sqrt{5}$, .009137306....., and $\sqrt{305}$.

Identify which group of real numbers each of the following belong. If a number belongs to more than one group, name the smallest group. (The number 57 is a counting number, a whole number, an integer, and a rational number, but the smallest group it belongs to is the counting numbers.)

1) 5 _____
2) $\sqrt{2}$ _____
3) 0 _____
4) .812.... _____
5) .416$\bar{9}$ _____
6) 2, 3, 4 _____
7) $\sqrt{21}$ _____
8) 9 _____
9) -1, 0, 1 _____
10) .0013 _____
11) -1 _____
12) $\frac{5}{6}$ _____

13) -2, -4, -6 _____
14) 8, 7, 6 _____
15) $\sqrt{7}$ _____
16) .75 _____
17) $5\frac{1}{4}$ _____
18) 0, 1, 2 _____
19) .0907... _____
20) $\frac{21}{23}$ _____
21) 5, 0, -5 _____
22) $\sqrt{11}$ _____
23) .217$\bar{8}$ _____
24) 6, 3, 0 _____

25) $\sqrt{34}$ _____
26) -2 _____
27) .43... _____
28) $\frac{8}{3}$ _____
29) .4812 _____
30) -1, -2 _____
31) .13$\bar{3}$ _____
32) 16, 17 _____
33) 0 _____
34) $\sqrt{16}$ _____
35) -2, 0, 2 _____
36) 3, 2, 1, 0 _____

Properties

Properties - are rules that tell you what you can and cannot do in Mathematics. Each step taken in solving an equation can be justified because one of the properties allows you to taked that step.

1. **Commutative property of addition and multiplication:** In addition and multiplication problems the terms can be moved without changing the value of the problem: 5+2 = 2+5; 2 • 3 • 4 = 4 • 2 • 3
2. **Associative property of addition and multipication:** In addition and multiplication it does not matter which terms you group together first: (6+3) +2 = 6 + (3+2); $(a \cdot b) \cdot c = a \cdot (b \cdot c)$
3. **Distributive property:** When there is a term next to a parentheses, and no sign between the term and the parentheses, every term inside the parentheses must be multiplied by the term outside the parentheses.   2(3x + 7) = 2(3x) + 2(7)
4. **Additive inverse:** The term you add to another term to get 0: 3a + (-3a) = 0

Identify the property that each statement represents:

1) 3 + 8 + 4 = 8 + 4 + 3

2) 3 • 8 • 4 = 8 • 4 • 3

3) a(2b - 7) = a(2b) + a(-7)

4) (+7) + (-7) = 0

5) 5 • (3x • 1) = (5 • 3x) • 1

6) 4 + (2 + 5) = (4 + 2) + 5

7) 11 • 2 • 5 = 2 • 5 • 11

8) 1 + 9 + 4 = 9 + 4 + 1

9) $a \cdot 6x \cdot y = 6x \cdot y \cdot a$

10) 8(9y-2) = 8(9y) +8(-2)

11) 9 + 7 + 3 = 7 + 3 + 9

12) 9 +(7 + 3)=(9 + 7) +3

13) (4xy) + (-4xy) = 0

14) a(2x+6) = a(2x) +a(+6)

15) 4 • 9 • 10 = 9 • 10 • 4

16) $x(yz) = (xy)z$

Properties continued:

17) x + (y + z) = (x + y) + z

18) 3z(2x + 5) = 3z(2x) + 3z(5)

19) a + c + x = c + x + a

20) $x \cdot y \cdot z = y \cdot z \cdot x$

21) $3y \cdot (a \cdot b) = (3y \cdot a) \cdot b$

22) (abx) + (-abx) = 0

23) 4(2x - 3) = 4(2x) + 4(-3)

24) $4 \cdot (12 \cdot 5) = (4 \cdot 12) \cdot 5$

25) $3a \cdot 2y \cdot b = 2y \cdot b \cdot 3a$

26) 2x + 3x + 7x = 3x + 7x + 2x

27) a + (2a + 3) = (a + 2a) + 3

28) $3 \cdot (2a \cdot 7a) = (3 \cdot 2a) \cdot 7a$

29) (-16x) + (+16x) = 0

30) -B + B = 0

31) (12 + 9) + 2 = 12 + (9 + 2)

32) $2 \cdot (3 \cdot 12) = (2 \cdot 3) \cdot 12$

33) x + 2x + 4x = 2x + 4x + x

34) 3a(y - z) = 3a(y) + 3a(-z)

35) $2x \cdot 3y \cdot 4 = 3y \cdot 4 \cdot 2x$

36) (y + 6y) + 5y = y + (6y + 5y)

37) 2 + 5a + 8 = 2 + 8 + 5a

38) $(3x \cdot 7b) \cdot 2 = 3x \cdot (7b \cdot 2)$

39) (175) + (-175) = 0

40) B(ig + ren) = Big + Bren

41) (7 + 10) + 34 = 7 + (10 + 34)

42) (Whit) + (-Whit) = 0

More Properties

5. **Multiplicative inverse:** The term you multiply another term by to get a product of 1, also called the reciprocal: $\frac{3}{7} \cdot (\frac{7}{3}) = 1$; $a \cdot \frac{1}{a} = 1$

6. **Identity property of addition:** The addition of zero to any term is equal to that term itself: $3x + 0 = 18$ and $27 + 0 = 27$
$$3x = 18$$

7. **Identity property of multiplication:** The multiplication of any term by 1 is equal to that term itself: $1a = 5$ and $1x = x$
$$a = 5$$

8. **Symmetric property:** The two sides of an equation can be interchanged:
$$15 = 2x - 9$$
$$2x - 9 = 15$$

Identify the property that belongs with each statement:

1) $\frac{1}{3} \cdot 3 = 1$
   _____

2) $25x + 0 = 25x$
   _____

3) $1y = y$
   _____

4) $7a = x$
   $x = 7a$
   _____

5) $0 + 11x = 33$
   $11x = 33$
   _____

6) $0 = x^2 + 8x + 9$
   $x^2 + 8x + 9 = 0$
   _____

7) $1x = 26$
   $x = 26$
   _____

8) $3x \cdot \frac{1}{3x} = 1$
   _____

9) $4x + 0 = 4x$
   _____

10) $x = y$
    $y = x$
    _____

11) $\frac{2}{5} \cdot (\frac{5}{2}) = 1$
    _____

12) $1 \cdot 18 = 18$
    _____

13) $x^2 - 16 = 0$
    $0 = x^2 - 16$
    _____

14) $(-10) \cdot (-\frac{1}{10}) = 1$
    _____

15) $0 + (-3Z) = 27$
    $-3Z = 27$
    _____

16) $1 \cdot x = x$
    _____

17) $15 = 7a$
    $7a = 15$
    _____

18) $3.6 = 1(3.6)$
    _____

19) $6 = 0 + 6$
    _____

20) $\frac{x}{a} \cdot \frac{a}{x} = 1$
    _____

21) $9x + 0 = 40$
    $9x = 40$
    _____

22) $a = 2c - 3$
    $2c - 3 = a$
    _____

23) $1z = 14$
    $z = 14$
    _____

24) $4\frac{1}{2} \cdot \frac{2}{9} = 1$
    _____

25) $56 = x$
    $x = 56$
    _____

26) $5a + 0 = 5a$
    _____

27) $6y = 1(6y)$
    _____

More Properties

9. Equality property: Whatever you do to one side of an equation you must do the same thing to the other side.

Examples

Equality Addition	Equality Multiplication	Equality Division
$2x - 9 = +4$	$2x = 16$	$6x = -42$
$+9 \quad +9$	$\frac{1}{2} \cdot 2x = 16 \cdot \frac{1}{2}$	$\frac{6x}{6} = \frac{-42}{6}$

10. Transitive property: (A) If one term is equal to a second term and it is also equal to a third term, the second and third terms will be equal. **(B)** Also, if one term is equal to a second term and the second term is equal to a third term, the first and third terms will be equal.

(A)	Examples	(B)
$6 = 1 + 5$		$x = 14$
$6 = 2 + 4$		$14 = c$
$1 + 5 = 2 + 4$		$x = c$

11. Substitution: If one term is equal to a second term, the second term can be substituted for the first term.   **Examples**

a. $3x = 26 - 6$
    $3x = 20$
    (20 is substituted for 26 - 6)

b. $x = (-7)$
    $5x + 12 = 22$
    $5(-7) + 12 = 22$ (-7 is substituted for x)

Identify the property that each statement represents:

1) $(7)2x = (7)28$

2) $\frac{5x}{5} = \frac{45}{5}$

3) $a = b$
   $a = c$
   $b = c$

4) $x + 26 = 40$
   $ - 26 = -26$

5) $x = 24 - 10$
   $x = 14$

6) $2b = 8 + 3$
   $8 + 3 = 11$
   $2b = 11$

7) $\frac{12a}{12} = \frac{132}{12}$

8) $32 = 20 + 12$
   $32 = 40 - 8$
   $20 + 12 = 40 - 8$

9) $a = 9$
   $4a - 6 = 3a + 3$
   $4(9) - 6 = 3(9) + 3$

10) $r = w$
    $r = t$
    $w = t$

11) $(\frac{1}{5})5x = (\frac{1}{5})65$

12) $8y - 2y = 21$
    $6y = 21$

13) $R = P - Q$
    $R = 22$
    $P - Q = 22$

14) $(\frac{1}{2})2x = (\frac{1}{2})8$

15) $a + 12 = -31$
    $ - 12 = -12$

16) $b = 4$
    $3b + 21 = 33$
    $3(4) + 21 = 33$

17) $\frac{21y}{21} = \frac{420}{21}$

Review of Properties

Identify the property that each statement represents:

1. $23 + (-23) = 0$
   _____

2. $5(3x - 2) = 5(3x) + 5(-2)$
   _____

3. $5x - 8 = 22$
   $ +8 +8$
   _____

4. $1(6x) = 6x$
   _____

5. $5 + (3 + 9) = (5 + 3) + 9$
   _____

6. $a = b$
   $b = 27$
   $a = 27$
   _____

7. $4x = 29 - 9$
   $4x = 20$
   _____

8. $\frac{1}{4}(4x) = \frac{1}{4}(36)$
   _____

9. $3 \bullet 5 \bullet 8 = 8 \bullet 3 \bullet 5$
   _____

10. $\frac{3}{7} \bullet (\frac{7}{3}) = 1$
    _____

11. $3y + 0 = 3y$
    _____

12. $\frac{8x}{8} = \frac{40}{8}$
    _____

13. $3 \bullet (4 \bullet 7) = (3 \bullet 4) \bullet 7$
    _____

14. $x = 6$
    $3x + 21 = R$
    $3(6) + 21 = R$
    _____

15. $22 = 3x - 7y$
    $3x - 7y = 22$
    _____

16. $Y = 10x$
    $Y = 50$
    $10x = 50$
    _____

17. $2x + 4x + 7x = 4x + 7x + 2x$
    _____

18. $8x + 14 = 22$
    $ -14 -14$
    _____

19. $-48x + 48x = 0$
    _____

20. $11y - 10y = 124 - 83$
    $y = 41$
    _____

Addition

When the signs of the numbers are alike: Add the numbers and keep the same sign.
Examples:
   a. $4 + 15 = +19$ b. $-8 - 11 = -19$ c. $+21 + 10 = 31$ d. $-6 - 4 = -10$

When the signs of the numbers are unlike: Subtract the absolute value of the numbers (the smallest number from the largest) then take the sign of the largest number.
Examples:
   a. $-32 + 14 = -18$ b. $+18 - 21 = -3$ c. $-9 + 15 = 6$ d. $+10 - 7 = 3$

Add the following problems:

1. $+3 + 8 =$
2. $-6 - 4 =$
3. $-3 + -8 =$
4. $4 + 1 =$
5. $-7 - 3 =$
6. $-4 - 11 =$
7. $+10 - 4 =$
8. $-15 + 9 =$
9. $+8 - 16 =$
10. $-9 + 20 =$
11. $+16 - 12 =$
12. $-9 + 5 =$
13. $-20 - 3 =$
14. $-15 - 2 - 6 =$
15. $14 + 2 + 9 =$
16. $-9 - 3 - 11 =$
17. $-16 + 11 =$
18. $+14 + 8 =$
19. $+23 - 2 =$
20. $+9 - 7 =$
21. $-2 + 17 =$

22. $-19 + 9 =$
23. $+14 - 8 =$
24. $-5 + 18 =$
25. $+6 + 18 =$
26. $+1 - 13 =$
27. $-3 + 15 =$
28. $-19 - 5 =$
29. $-10 + 19 =$
30. $-15 + 8 =$
31. $+4 - 9 =$
32. $(-6) + (-7) =$
33. $21 + (-15) =$
34. $(+9) + (+14) =$
35. $-8 + (-3) =$
36. $(+8) + 5 =$
37. $(-13) + (-4) =$
38. $8 + (+7) + 2 =$
39. $-7 + (-4) - 10 =$
40. $+(11) + 10 =$
41. $-8 + (-5) + (6) =$
42. $(-4) + (-5) + (-9) =$

43. $-4 + (-11) =$
44. $(-6) - 14 =$
45. $-7 + 13 =$
46. $+18 - 17 =$
47. $-11 + 12 =$
48. $-18 - 6 =$
49. $+5 - 5 =$
50. $-20 - 4 =$
51. $+19 - 12 =$
52. $-14 + 6 =$
53. $-3 - 7 =$
54. $+7 - 11 =$
55. $+4 + 16 =$
56. $+15 - 15 =$
57. $-12 + 10 =$
58. $-16 - 1 =$
59. $+6 - 9 =$
60. $+20 - 10 =$
61. $-8 + 8 =$
62. $+16 + (20) =$
63. $(-7) + (+10) =$

64. (+11) + (+3) =
65. 7 + (-2) =
66. -10 + (+4) =
67. (-9) + (-7) =
68. +10 + (-14) =
69. (-8) + (-15) =
70. (+9) + (-18) =
71. (-6) + (+7) =
72. (28) - 13 =
73. -8 + (-19) =
74. (-17) - 21 =
75. 8 + (-15) =
76. (-13) -8 =
77. (+12) + (-19) =
78. -13 + (+24) =
79. (-18) + 13 =
80. +18 + (-14) =
81. (-32) + (-14) =
82. (-4) +24 =
83. 9 + 9 + 3 =
84. 3 +19 + -22 =
85. +13 + -5 -6 =
86. +11 +9 -31 =
87. -4 -6 -5 =

88. -17 + -1 +-7 =
89. -9 + (-7) =
90. (-3) + (+5) =
91. (+21) - 14 =
92. -30 + (+12) =
93. -6 + (-9) =
94. +13 +(-2) =
95. +15 + -30
96. -16 + (+3) =
97. -8 + (-21) =
98. +5 + (-11) =
99. (+2) -6 =
100. +17 + (+12) =
101. +21 + (-13) =
102. -7 + (-17) =
103. -19 + (+13) =
104. (-21) + (32) =
105. -18 - 4 =
106. (+14) + (-19) =
107. (+22) - 14
108. (21) + (+12) - 13 =
109. -7 + (+6) =
110. (-6) + 17 =
111. 15 + (8) =

112. (-3) + (10) =
113. -5 - 6 + (-3) =
114. (-3) + 14 =
115. (-16) + (-3) =
116. (-5) + 4 =
117. -16 + (18) + (-2) =
118. 7 + (-9) - 3 =
119. 14 +(-6) + 1 =
120. -14 +(-20) +7 =
121. 15 +11 -6 =
122. (5) +(-8) + (3) =
123. (-10) -7 +(-5) =
124. (8) -3 +(+2) =
125. 6 - 21 + 9 =
126. -18 + 7 - 6 =
127. (-9) -5 =
128. (14) + (+4) =
129. (-8) + (-4) =
130. (+5) + (7) =
131. (+7) - 13 =
132. (-7) -10 =
133. -7 + (-6) =
134. 20 + (-15) + -7 =
135. -16 - 8 - 10 =

Subtraction

1. There are very few subtraction problems in algebra. Most problems that look like subtraction can be <u>changed to addition</u> by putting a positive (+) sign in front of the negative (-) sign.

    Examples: -12 - 6   and   -11 (can be   -11
    (can be changed to) -12 + (-6)   - 3 changed to) +(-3)

2. Two types of problems cannot be changed to addition by putting a positive sign in front of a negative sign: **a. Those that specifically say "subtract", and b. Those that have a negative sign in front of a parentheses.**

3. The rule for subtraction is: **change the sign (or signs) of the terms you are subtracting by, then use the rules for addition.** Subtraction is often referred to as "adding the additive inverse". In examples (a), (b), and (c), the botton sign is circled and replaced by the opposite sign.

Examples:

a. Subtract: b. Subtract: c. Subtract: d. (-19) - (-25)   e. +14 - (-10)
   +21   - 37   - 45   -19 + 25 = 6   +14 + 10 = +24
   ⊕13   ⊕20   ⊖22
   +34   - 17   -67

Subtract:

1. +15   2. -14   3. +21   4. -11   5. +14   6. -14   7. 23   8. +16
   -6   - 3   -30   -21   - 3   +21   -40   +19

9. +18   10. -32   11. +17   12. -15   13. +18   14. -3   15. -25   16. -8
   +10   +18   +25   +24   -34   +7   -39   +23

Simplify: 17. +13 - (+24)   18. -2 - (+19)   19. 15 - (-2)

20. -13 - (-15)   21. +34 - (+45)   22. +16 - (+9)   23. -29 - (-17)

24. -13 - (+31)   25. 26 - (-18)   26. +35 - (+49)   27. -53 - (+24)

28. +36 - (-41)   29. -3 - (-1)   30. +55 - (-32)   31. 28 - (63)

32. -28 - (+54)   33. +40 - (-19)   34. -29 - (+43)   35. 17 - (-9)

Multiplication

1. When multiplying **two** terms, **a.** if one term is negative, the product is negative. **b.** If both terms are **either** positive or negative, the product is positive. (examples a,b,c,d)

2. When multiplying more than two terms, an **odd** number of negative signs produces a negative product, an **even** number of negative signs produces a positive product. (examples e,f,g,h)

Examples

a. $(-4)(3) = -12$ b. $+4(-9) = -36$ c. $-6(-2) = +12$ d. $(+7)(+2) = +14$

e. $(-6)(-2)(5) = +60$ f. $(-3)(2)(+4) = -24$ g. $-1(-7)(-3) = -21$ h. $-5(3)(-6)(+2) = 180$

Multiply:

1. $(-6)(3) =$
2. $-4(-2) =$
3. $(-7)(-3) =$
4. $(+6)(3) =$
5. $+7(-3) =$
6. $(-9)(-6) =$
7. $(4)(-2) =$
8. $-3(+8) =$
9. $-5(-7) =$
10. $(-4)(4) =$
11. $+6(+2) =$
12. $-3(+3) =$

13. $-9(-8) =$
14. $(8)(+3) =$
15. $(-4)(2) =$
16. $(-6)(-4)(3) =$
17. $+7(3)(-2) =$
18. $-9(-3)(-1) =$
19. $(6)(4)(+2) =$
20. $(2)(-3)(+7) =$
21. $(7)(-1)(-2) =$
22. $(-6)(+4)(-3) =$
23. $-5(-7)(-1) =$
24. $6(+5)(-3) =$

25. $-5(-2)(+5) =$
26. $(7)(+6)(-2) =$
27. $(-6)(-3)(-5) =$
28. $(+7)(-4)(3) =$
29. $(+2)(+9)(4) =$
30. $(-9)(3)(-6) =$
31. $+7(+7)(-1) =$
32. $-3(+6)(-5) =$
33. $+4(-4)(3) =$
34. $-10(-9)(-5) =$
35. $(-6)(-3)(2) =$
36. $+7(2)(-\frac{1}{2}) =$

37. $(2)(-4)(-\frac{1}{4}) =$
38. $-2(-3)(5)(-2) =$
39. $(+4)(5)(\frac{1}{5}) =$
40. $-4(4)(-2)(-1) =$
41. $(-3)(-3)(2) =$
42. $(-5)(6)(\frac{1}{10}) =$
43. $3(4)(-2)(\frac{1}{3}) =$
44. $(+7)(6)(3)(-2) =$
45. $-5(4)(-2)(\frac{2}{5}) =$
46. $(2)(-3)(-5)\left(-\frac{2}{3}\right) =$
47. $6(-6)(2)(3) =$
48. $(-2)(\frac{1}{2})(4)(-3) =$

11

Division

The same rules used in multiplication are used in division. **a.** If one term is negative, the quotient is negative. **b.** If both terms are either positive or negative, the quotient is positive.

Examples

a. $\frac{-15}{+3} = -5$

b. $\frac{-24}{-6} = +4$

c. $-4\overline{)-28} = +7$

d. $-6\overline{)48} = -8$

e. $(+35) \div (-7) = -5$

f. $(-40) \div (-4) = +10$

Divide:

1. $\frac{-42}{-21} =$
2. $\frac{+30}{-3} =$
3. $\frac{-32}{+8} =$
4. $\frac{20}{-5} =$
5. $\frac{-14}{-2} =$

6. $\frac{-25}{5} =$
7. $\frac{+24}{+8} =$
8. $\frac{-39}{13} =$
9. $\frac{-36}{-3} =$
10. $\frac{28}{-7} =$

11. $\frac{-22}{+11} =$
12. $\frac{+18}{3} =$
13. $\frac{-26}{-13} =$
14. $\frac{-7}{+7} =$
15. $\frac{45}{-9} =$

16. $+30 \div (-6) =$
17. $(-16) \div (-2) =$
18. $(+19) \div (-1) =$

19. $-20 \div (+4) =$
20. $(-32) \div (16) =$
21. $40 \div (-20) =$

22. $(28) \div (+14) =$
23. $-26 \div (-2) =$
24. $(-10) \div 5 =$

25. $+45 \div (-9) =$
26. $-24 \div (-3) =$
27. $(+9) \div (+3) =$

28. $(-15) \div (+3) =$
29. $+40 \div (-20) =$
30. $(+12) \div -3 =$

31. $-4\overline{)20}$
32. $+11\overline{)-33}$
33. $-6\overline{)-18}$
34. $+8\overline{)+16}$

35. $+7\overline{)-21}$
36. $-16\overline{)32}$
37. $-10\overline{)90}$
38. $9\overline{)-36}$

39. $-12\overline{)-24}$
40. $20\overline{)+40}$
41. $+3\overline{)-27}$
42. $-30\overline{)60}$

43. $-9\overline{)-18}$
44. $25\overline{)-75}$
45. $-8\overline{)+40}$
46. $-11\overline{)44}$

Order of operation

In most algebra problems, you will have to do more than one operation, (add, subtract, multiply, divide). You need to know which of these to do first, which to do second, and so on. This is called the **order of operation.** First you must know that if there are no exponents, you should multiply and divide (in the order that they appear in the problem) **before** you add and subtract.

Examples

a. (-3)(-4) + 2(-5)
    = +12 + (-10)
    = +2

b. (-6)(-3) - 2(-5) + (5)(-4)
    = +18   +10   + (-20)
    = +28   -20
    = +8

Simplify using the correct order of operation:

1. (5)(-4) + (-3)(-2) =

2. (-2)(-5) - (6)(-2) =

3. (3)(+4) + (5)(+4) =

4. (-2)(5) - (+3)(-3) =

5. (-1)(-8) - (+9)(-5) =

6. (4)(-6) + (+8)(+2) =

7. (+5)(5) + (-4)(-8) =

8. (+6)(-4) - (-7)(+4) =

9. (-7)(-3) + (-6)(6) =

10. (-2)(7) - (-5)(-3) =

11. (+9)(-4) + (-8)(5) =

12. (-10)(-2) - (9)(-4) =

13. (-2)(4) + (5)(+2) =

14. (-5)(-2) - (-3)(-6) - (-4)(-3) =

15. (7)(-3) + (-4)(-6) - (2)(-10) =

16. 3(+5) - 6(4) + 5(-6) =

17. -7(-2) + 5(3) - 4(10) =

18. (4)(-3)(+2) - (-6)(4)(-2) =

19. 3(-2)(-5) + 7(2)(-1) =

20. (3)(-3)(-9) - (-6)(-2)(-8) =

Order of Operation - Exponents

When there are exponents in the problem, remove the exponents by multiplying, then use the rule for order of operation: multiply, divide before add, subtract.

Examples

a. $4 + 5^2 \cdot 3$
   $= 4 + 25 \cdot 3$
   $= 4 + 75$
   $= 79$

b. $5 \cdot 4^2 + 8 - 3^2 \div 3 \cdot 6$
   $= 5 \cdot 16 + 8 - 9 \div 3 \cdot 6$
   $= 80 + 8 - 3 \cdot 6$
   $= 80 + 8 - 18$
   $= 88 - 18$
   $= 70$

Simplify using the correct order of operation:

1. $2 + 3^2 \cdot 5$
2. $8^2 - 3 \cdot 5$
3. $7 + 5 \cdot 4 \div (-10)$

4. $4 + 5^2 \cdot 2$
5. $4 + 7 \cdot 3^2$
6. $3 \cdot 8 + 36 \div 3^2$

7. $10 + 4^2 \cdot (-3)$
8. $8 \times 4 \div 2^2$
9. $5 \cdot 3 \cdot 24 \div 2^2$

10. $6 - 9 \cdot 3^2$
11. $6 \cdot 5 \cdot 3^2$
12. $4^2 \cdot 2 \div 15 + 5$

13. $8 + 5^2 \cdot 6$
14. $15 \div 3 + 2^2$
15. $8 \cdot 3^2 \div 24 - 6$

16. $5^2 + 9 \cdot 3$
17. $6 - 2 \cdot 8 \div 4^2$
18. $4 \cdot 9 \div 3^2 + 2^2$

19. $5 + 4 \cdot 2^2$
20. $9 \div 3^2 + 5$
21. $8 \div 2^2 + 6 \cdot 3$

Order of Operation - Removing Parentheses

Removing parentheses must be done in all types of algebra problems. There are three possible situations in which parentheses can be removed.
1. When there is a **positive** sign in front of the parentheses, nothing changes when the parentheses are removed. (see example 1.)
2. When there is a **negative** sign in front of the parentheses, all the signs inside the parenthese are **changed** when the parentheses are removed. (see example 2.)
3. When there is a term **next to** a parentheses, and **nothing between** the term and the parentheses, everything inside the parentheses is multiplied by the term outside the parentheses. (see example 3.)

Examples

1. $6 + (4 - 3)$
   $6 + 4 - 3$
   $10 - 3 = 7$

2. $6 - (9x - 2)$
   $6 - 9x + 2$
   $8 - 9x$

3. $4(8x - 3)$
   $4(8x) + 4(-3)$
   $32x - 12$

Remove the parentheses and simplify:

1. $7 + (9 - 3)$
2. $4 - (6 - 5x)$
3. $8(2 + 7)$
4. $-6(8y - 3)$
5. $11y + (3 + 10)$
6. $7 - (4a + 2)$
7. $6(5 - 2)$
8. $-9(48 + 5)$
9. $-8 + (-3 + 2)$
10. $-4 - (10 - 3)$
11. $9(-5 + 8)$
12. $-7(-3x - 8)$
13. $-10 + (-3 - 12)$
14. $-4x - (-6 - 2)$
15. $10(-6 - 8)$
16. $-4(-5 - 9)$
17. $5 + (-11 + 2)$
18. $+10 - (-8 + 3)$
19. $-2 - (8y + 2)$
20. $+6(6 - 11)$
21. $-4 + (4 - 3)$
22. $+7 - (7 - 2)$
23. $8(-8r - 3)$
24. $-3(6 - 1)$
25. $-5 + (5 - 2)$
26. $6x + (2x - 3)$
27. $-4 - (4 + 2a)$
28. $-2 + (4y - 3)$
29. $+3(2 - 10b)$
30. $+7(-8 - 4)$
31. $-3 - (3b - 6)$
32. $5 + (6 - 2y)$
33. $8 - (-8 + 7)$
34. $-7(2y - 4)$
35. $+2 - (4 - 2x)$
36. $10(4 - 7)$

Order of Operation - Parentheses Inside Parentheses

Often you will find parentheses **inside** of parentheses, brackets, and braces. The rule for order of operation states that to remove the parentheses, brackets, and braces, **begin by simplifying what is inside the innermost** parentheses, bracket, or brace. Next remove that innermost parentheses, bracket, or brace. Repeat the process with the next innermost parentheses, bracket, or brace.

Examples

a. $1 - (11 + [6 - 3] - 8)$
    $1 - (11 + [3] - 8)$
    $1 - (11 + 3 - 8)$
    $1 - (6)$
    $1 - 6$
    -5

b. $8 + (4 - 5[6 - 2] + 9) - 3$
    $8 + (4 - 5[4] + 9) - 3$
    $8 + (4 - 20 + 9) - 3$
    $8 + (-16 + 9) - 3$
    $8 + (-7) - 3$
    $8 - 7 - 3$
    -2

Simplify using the correct order of operation:

1. $6 + [8 + 2] - 4$

2. $9 - (3 + 2) + 4$

3. $7 + [6 - 2] - 11$

4. $3 - (8 - 4) + 5$

5. $(11 + [-7 + 2] - 3)$

6. $(1 + 4(2 + 1) - 6)$

7. $[8 + 7(9 - 3) + 7]$

8. $(10 - 3[10 + 2] + 20)$

9. $(7 - 4[7 - 2] - 10)$

10. $6 - 5(4 - 2) + 8$

11. $10 + 6(-3 - 5) - 7$

12. $15 + [+10 - 15] - 6$

13. $13 - [-3 + 2] + 8$

14. $11 + [9 - 5] + 9$

15. $(8 - 2[3^2 + 4] - 7)$

16

Order of Operation - Parentheses Inside Parentheses Continued

16. [12 - 10(2 - 1) + 6]

17. 2^2 + (8 + [5 - 4] - 12)

18. 7 + (3 - [8 - 5] - 9) + 1

19. 10 + 2(9 + [3 + 1] + 8) + 8

20. 6 - 3(3 - [2 - 3] - 10) - 3

21. -5 + 4(9 - 3[9 + 2] - 4) + 9

22. 4 + (2 + 2[6 - 1] + 9) - 7

23. -9 - 7(7 - 5[4 - 7] + 1) + 10

24. 4 - (6 + [9 + 1] + 3)

25. 5 - (13 + 4[7 - 1] + 8) + 2

26. 9 + (3 + [9 + 1] - 11) - 5

27. 10 + [8 - 4(6 - 4) + 9] + 7

28. 7 - 2(7 - [5 + 2] + 6) - 3

29. 9 + 3(10 + [8 - 4] - 11) + 2

30. 15 - 5(7 + 2[3^2 - 2] + 19) + 1

31. 11 + 4(3 - [2^2 + 2] - 4) + 12

32. 7^2 - [9 - 2{6 - 3 + 2} - 6] + 2

Order of Operation - Substitution

In Algebra the solution to an equation can be substituted into the equation to check the correctness of the equation. After the substitution has been made, all the rules for order of operation must be followed.

Examples

a. Is (-4) the solution to the equation:
$$7x - 18 = 5x - 26 \ ?$$
$$7(-4) - 18 = 5(-4) - 26$$
$$-28 - 18 = -20 - 26$$
$$-46 = -46$$
Since the left side of the equation equals the right side, (-4) is the correct solution

b. Is (+3) the solution to the equation:
$$2x^2 - 9x + 9 = 0 \ ?$$
$$2(3)^2 - 9(3) + 9 = 0$$
$$2(9) - 27 + 9 = 0$$
$$18 - 27 + 9 = 0$$
$$-9 + 9 = 0$$
$$0 = 0 \ \textbf{yes}$$

By substitution, determine if the numbers inside the parentheses are the solutions to the following equations:

1. $x + 5 = 11, (x = 6)$

2. $x - 7 = -4, (x = 3)$

3. $\frac{x}{7} = 3, (x = 21)$

4. $6x = 18, (x = 3)$

5. $\frac{2x}{3} = 8, (x = 12)$

6. $4x + 1 = 7, (x = 1\frac{1}{2})$

7. $6x - 2 = 30, (x = 6)$

8. $\frac{x}{5} - 2 = 4, (x = 35)$

9. $\frac{x}{3} + 7 = -2, (x = -27)$

10. $3x + 1 = 2x - 4, (x = -5)$

11. $6x - 4 = 5x + 9, (x = 13)$

12. $8x + 10 = 3x + 50, (x = 10)$

13. $5x - 4 = x - 32, (x = -7)$

14. $7x + 3 = 4x + 21, (x = 5)$

15. $x^2 = 81, (x = 6)$

16. $2x^2 = 32, (x = -4)$

17. $x^2 - 13x + 30 = 0, (x = 10)$

18. $x^2 + 8x - 20 = 0, (x = -2)$

Order of Operation - Substitution

In Algebra, the solution to formulas can be determined by substituting the values of the variables into the formulas. After the substitutions have been made, all the rules for order of operation must be followed.

Examples

a. In the following formula, find c when $x = 7$: $c = x^2 - 3x + 2$

$$c = (7)^2 - 3(7) + 2$$
$$c = 49 - 21 + 2$$
$$c = 28 + 2$$
$$c = 30$$

b. In the following formula, find R when $a = -2$ and $b = 5$

$$R = \frac{4a-3b}{-6ab}$$
$$R = \frac{4(-2)-3(5)}{-6(-2)(5)}$$
$$R = \frac{-8-15}{60} = \frac{-23}{60}$$

By substitution, find the values of the following formulas:

1. $(x = 3), b = 2x^2 - 5$

2. $(x = 4), a = 3x^2 + 7$

3. $(x = 6), c = x^2 + 17$

4. $(x = -3), r = 2x^2 + 10$

5. $(x = -5), t = 5x^2 - 11$

6. $(x = 2), y = 4x^2 + 7$

7. $(x = -6), q = 3x^2 - 20$

8. $(x = 10), r = 7x^2 + 17$

9. $(x = -8), p = 4x^2 - 8$

10. $(x = 2, y = 3), b = \frac{xy}{-3}$

11. $(a = 2, b = 7), x = \frac{3ab}{14}$

12. $(p = 6, q = -1), t = \frac{p+q}{5}$

13. $(b = -6, c = 5), r = \frac{3bc}{-2}$

14. $(y = 9, a = -6), q = \frac{4ya}{12}$

15. $(r = -3, b = -2), y = \frac{rb-5}{-1}$

16. $(t = 2, q = -7), y = \frac{3qt-3}{20}$

17. $(x = -5, z = 3), a = \frac{4xz-10}{-7}$

18. $(b = 4, r = 6), c = \frac{3br-12}{4}$

Review of Order of Operation

Simplify the following mathematical expressions:

1. (-2)(+4) + (-8)(+4)

2. $3^2 + 12 \div 2$

3. 4 - (4x + 7)

4. 7(3-6) - 10

5. Check the equation:
   $x^2 + 8x - 3 = 62$; (x = 5)

6. Find the value of P:
   P = 29 - 3y; (y = -6)

7. 9 + (2z + 17z)

8. 3 - (4+[6-1] - 5)

9. $10 - 3 \bullet 4 + 3^2$

10. 20x - (11x - 32)

11. $5^2 - 20 \div 4 \bullet 2^2$

12. 5(9 +[3+2]-9)-1

13. Check the equation:
    $\frac{3x}{5} - 2 = 28$; (x = -10)

14. 9x - (24 - x)

15. 9 +(4 - 5[9- 4] -8) + 16

16. 3(13x - 2)

17. Find the value of V:
    $V = 2p^2 + 4p$; (p = 7)

18. $1 + 24 \div 3 - 4^2 \bullet 3$

19. Check the equation:
    3x - 2 = 5x + 8; (x = -5)

20. 5 -[4 - 2(11-5) +7] +9

21. Find the value of J:
    $J = Q^2 + 3Q$; (Q = -10)

Chapter Review
Name the underlined item:

1. 5w$\underline{^2}$ _____

2. $\underline{7x + 12}$ _____

3. 4y$\underline{z}^2$ _____

4. $\underline{-}$2x _____

5. $\underline{5}$cd _____

Identify the number:

6. $7\frac{1}{2}$ _____

7. $\sqrt{17}$ _____

8. .375 _____

9. 43 _____

10. -9 _____

Name the property:

11. 3z + (-3z) = 0 _____

12. 4 • 7 • 3 = 3 • 7 • 4 _____
_____

13. 7 + 2 + 5 = 5 + 7 + 2 _____
_____

14. 1y = y _____

15. z + 3 = 7
       - 3 -3 _____
_____

16. $2 \times \frac{1}{2} = 1$ _____

17. $\frac{6x}{6} = \frac{42}{6}$ _____

18. p = 9
    q = 9
    p = q _____

19. 3(4x - 3) = 12x - 9 _____

20. 9 = 5x + 4
    5x + 4 = 9 _____

21. (5 + 7) + 4 = 5 + (7 + 4) _____
_____

22. $\frac{1}{4}(4x) = \frac{1}{4}(8)$ _____

23. z = 7
    -3x - 12 = -33
    -3(7) - 12 = -33 _____

24. q + 0 = q _____

25. 4(6 + 3) = 4(9) _____

Simplify:

26. (-2) + 4

27. 5(2x - 3)

28. 10 + (2y - 12)

29. -5 - (4p - 9)

Chapter 1 Review Continued

30. $4^2 - 5 \bullet 3 + 3^2$

31. $17 - 2^2 + 3 \bullet 9 - 5^2$

32. $(+6)(-3) - (4)(-7)$

33. $(-4)(-3) + (-2)(-5) - (-6)(-7)$

34. $5 + (6 - 2[4-7] + 3) - 10$

35. $10 - (3 + 2\{5-2\} - 4) - 2$

36. Is "3" the solution to: $2x - 7 = -3$?

37. Is "20" the solution to: $\frac{3}{4}x - 3 = 12$?

38. Find A: $A = 2x^2 - 8$ when $x = 6$

39. Find P: $P = 4x^2 - 3x - 1$ when $x = 1$

Chapter 2
Addition and Subtraction of Polynomials

Addition of Polynomials - To add polynomials, add the numerical coefficients of the like terms. (Like terms are terms that have the **same** variables with the **same** exponents.) These problems are added **horizontally**.

Examples

a. $(5x + 3y) + (7x - 2y) = 12x + y$
  Add 5x and 7x to get 12x
  Add 3y and (-2y) to get y (1y)

b. $6a^2 + 8a + 3 + 4a - 11 + 2a^2 = 8a^2 + 12a - 8$
  Add $6a^2$ and $2a^2$ to get $8a^2$
  Add 8a and 4a to get 12a
  Add 3 and (-11) to get -8

Add the following problems by combining the like terms:

1. $(3x + 2) + (-2x + 4)$

2. $3a + 2 + 4a - 6$

3. $(7x - 4) + (4x - 9)$

4. $6x - 5 - 2x + 7$

5. $(4y + 2) + (4y - 9)$

6. $4y + 3 + 2y + 9$

7. $(2y - 9) + (-4y - 3)$

8. $8x - 6 - 10x - 5$

9. $(2a^2 - 5a) + (4a^2 + 7a)$

10. $3a - 5b + 6b - 2a$

11. $(6b^2 + 9b) + (-4b^2 - 4b)$

12. $4x + 7x - 5y - 3x$

13. $(3x^2 + 8x - 9) + (5x^2 - 10x + 4)$

14. $8a - 2b + 7b - 3b$

15. $(6x^2 - 3x + 7) + (-8x^2 - 4x + 1)$

16. $6y + 2z - 5z + 2y$

17. $(4x^2 - 7x - 3) + (5x^2 + 6x - 1)$

18. $x^2 + 8x - 3 + 4x^2 - 7x + 9$

Addition of Polynomials Continued:

19. $(7a^2 - 4ab - 9b^2) + (7a^2 - 3ab + 9b^2)$

20. $4x^2 - 6x - 5 - 2x^2 + 7x + 9$

21. $(8x^2 - 11x - 2) + (3x^2 - 2x + 10)$

22. $4a^2 + 3a + 9 + 5a^2 - 4a - 10$

23. $(18x^2 - 2x + 9) + (10x^2 - 5x - 7)$

24. $7x^2 - 3x + 4x^2 + 9x - 2x^2 + 7$

25. $(3a^2 + 4a + 2) + (6a^2 + 4a + 7)$

26. $8a^2 + 4a + 9a^2 - 3a + 2a^2 - 6a$

27. $(3x - 1) + (4x - 3) + (7x + 2)$

28. $10x^2 - 7x^2 - 4x + 3x + 7 - 2$

29. $(3a + 5) + (11a - 4) + (-10a - 1)$

30. $8y^2 + 7y^2 + 3y - 6y - 11 - 4$

31. $(6x - 3) + (4x + 7) + (-4x + 3)$

32. $6x^2 - 4x - 3 + 8x^2 - 9 + 2x^2 + 1$

33. $(a - 2b + 3c) + (6a + 4c) + (2a - 3b)$

34. $3a^2 + 5a + 2 - 4a + 7 - 3a - 2$

35. $(7b^2 - c) + (-4b^2 + 2c) + (3b^2 - 5c)$

36. $6y^2 - 7y + 8y^2 - 2y - 15 + 4y$

37. $(6x^2 - 4x + 9) + (x^2 - x) + (3x - 7)$

38. $9a^2 - 6a + 2a^2 + 9 + 5a - 2$

39. $(4a^2 - 3a) + (2a - 11) + (3a^2 + 20)$

40. $4x^2 - 3 + 2x^2 + 9 - 4x^2 + 7x$

41. $(6a - 3b) + (4a + 7c) + (9b - 4c)$

42. $3x^3 + 4x - 8 - 8x^3 + 7x^2 + 1 - 6x^2 + 5x^3$

43. $(9x^2 - 7xy + 6y^2) + (2x^2 + xy - 3y^2)$

44. $9ab - 3bc - 5cd - 2bc + ab - 12cd$

45. $(3a^2 + 4ab - 9b^2) + (a^2 - ab - b^2)$

46. $5x^2 + 8x^2 - 5x + 8x + 5 - 8$

Addition Problems Written Vertically - Polynomials can also be added when they are written vertically. The same rule applies: add the numerical coefficients of the like terms.

Examples

a. 23x - 12y
    14x +19y
    37x + 7y

b. $6y^2$ +8y - 17
    $4y^2$ - 9y - 8
    $-5y^2 - 7y + 22$
    $5y^2$ - 8y - 3

Add:

1. 3x - 2y
    2x + 7y

2. 13a -10b
    7a - 4b

3. $8y^2 + 2y$
    $4y^2 + 7y$

4. 6a - 4
    -5a + 7

5. 7y - 6
    4y + 9

4. $8a^2 - 2a$
    $-5a^2 - 7a$

7. $3x^2 - 8x + 9$
    $2x^2 + 7x - 3$

8. $9y^2 + 9y - 3$
    $4y^2 + 7y - 6$

9. $7b^2 - 4bc + 7c^2$
    $-6b^2 + 9bc - 11c^2$

10. $8x^2 - 4x - 3$
    $2x^2 + 7x - 10$

11. $9y^2$ -3
    $-5y^2 - 4y + 9$

12. 6a - 2
    -5a - 3
    7a + 6

13. 3y - 2z
    8y + 7z
    -2y + 3z

14. 9b - c
    -4b + 2c
    -3b - 6c

15. - 6x - 5
    - 3x + 2
    - 7x - 4

16. $5x^2 - 4x - 1$
    $3x^2 + 9x - 5$
    $-2x^2 + 7x + 2$

17. $-4y^2 - 2xy$
    $6y^2 + 9xy$
    $-5y^2 + 3xy$

18. $3x^3 - 4x^2 + 9x - 5$
    $-2x^3 + 6x^2 - 7x + 2$

19. $2a^3 - 5a^2b + 6ab^2 + 9b^3$
    $-7a^3 + 6a^2b - 4ab^2 + 9b^3$

20. $6y^2 + 8yz - 6z^2$
    $-10y^2 - 9yz + 7z^2$

21. $9a^2 - 4ab + 7b^2$
    $6a^2 - 9ab - 9b^2$

Subtraction of Polynomials when Written Horizontally - All horizontal subtraction problems will have a negative sign in front of a parentheses, which indicates the problem is subtraction. In one step make these three changes: **1) change all the signs <u>inside</u> the parentheses <u>following</u> the negative sign, 2) remove the negative sign in front of that parenthese, 3) remove all parentheses.** The last step is to combine the like terms as you would in an addition problem.

Examples

a. $(13x - 9y) - (5x + 7y)$
   $13x - 9y \ - 5x \ - 7y$
   $ 8x \ - 16y$

b. $(17x^2 - 4x + 7) - (10x^2 + 8x - 14)$
   $17x^2 - 4x + 7 \ - 10x^2 - 8x + 14$
   $ 7x^2 - 12x + 21$

Subtract these problems written horizontally:

1. $(4x - 6) - (3x + 1)$

2. $(4x^2 + 3x - 1) - (2x^2 + 4x - 1)$

3. $(8x + 3) - (3x - 5)$

4. $(7x^2 - 4x - 7) - (5x^2 + 2x + 7)$

5. $(2x - 7) - (7x - 6)$

6. $(20x^2 + 9x + 2) - (11x^2 - 3x - 11)$

7. $(9x - 9) - (2x + 4)$

8. $(17x^2 - 6x + 11) - (4x^2 - 7x + 6)$

9. $(12x + 11) - (9x + 9)$

10. $(13x^2 + 4x - 3) - (8x^2 + 4x + 1)$

11. $(11x - 3) - (5x - 7)$

12. $(9x^2 - 5x + 7) - (4x^2 - 5x + 3)$

13. $(5x + 1) - (x + 13)$

14. $(11x^2 + 13x - 2) - (10x^2 + 6x + 9)$

15. $(5x - 4) - (4x + 14)$

16. $(6x^2 + 4x + 9) - (3x^2 - 11x - 3)$

17. $(17x - 7) - (x - 1)$

18. $(4x^2 - 3x - 7) - (x^2 + 4x - 6)$

19. $(x + 10) - (5x + 3)$

20. $(18x^2 + 8x - 10) - (14x^2 - 7x - 12)$

Subtraction of Polynomials Continued:

21. $(12x - 3y) - (8x - 2y)$

22. $(8a^2 + 6a - 3) - (2a^2 + 4a)$

23. $(5y + 2z) - (2y + 4z)$

24. $(7y^2 - 6y + 2) - (y^2 - 7)$

25. $(3a - 8c) - (a + 9c)$

26. $(4y^2 + 9y) - (3y^2 - 7y + 6)$

27. $(4z - 6b) - (6z - 6b)$

28. $(8x^2 + 11) - (2x^2 + 6x - 3)$

29. $(9c + 4a) - (4c - 7a)$

30. $(7a^2 + 6) - (4a - 3)$

31. $(6x^2 + 8x - 1) - (2x^2 - 7x - 3)$

32. $(9x^2 - 16x + 4) - (10x + 4)$

33. $(2x^2 - 7x - 3) - (4x^2 - x + 5)$

34. $(3x^3 - 5x^2 + 2x - 1) - (x^3 - 2x^2 + 7x + 3)$

35. $(4x^2 + 9x + 2) - (7x^2 - 10x + 3)$

36. $(6x^3 + 10x^2 - 4x + 9) - (4x^3 + 7x^2 - 3x + 8)$

37. $(14x^2 + 4x + 3) - (10x^2 + 4x - 3)$

38. $(12x^3 + 9x^2 - 7x - 1) - (5x^3 - 16)$

39. $(5x^2 + 9x - 2) - (5x^2 - 8x + 2)$

40. $(8x^3 - 2x + 1) - (5x^3 + 6x^2 + 9)$

41. $(3a - 4b + 5c + d) - (a + b - c - 8d)$

42. $(2x^4 - 7x^2 + 4) - (3x^3 - 10x^2 + 15)$

43. $(9yz + 21az + 16ay) - (3yz + 2ay - 7az)$

44. $(x^2 - 4x + 19) - (-3x^2 + 8x - 5)$

45. $(a - x + b + y) - (a + x + b - y)$

46. $(2a^2 + 15ab + 3b^2) - (a^2 - 13ab + 19b^2)$

47. $(23a^2 - 32) - (17a^2 - 11a + 17)$

48. $(15x^3 + 3x - 1) - (6x^3 - 8x^2 + 27)$

Subtraction Problems Written Vertically - Polynomials can also be subtracted when they are written vertically. The instruction **Subtract** must precede the problem. Two steps must be taken: **1)** Change the sign of the bottom terms, (circle the old term and put in the new sign), then **2)** use the rules for addition.

Examples

These are the problems:

Subtract:
a. $7x - 3$
   $+ 2x - 4$

b. $7x^2 - 3x + 11$
   $+ 3x^2 + 6x - 15$

c. $6a^2 -6$
   $+ 4a^2 - 7a + 3$

These are the same problems with the signs changed:

$7x - 3$
$\ominus 2x \oplus 4$
$5x + 1$

$7x^2 - 3x + 11$
$\ominus 3x^2 \ominus 6x \oplus 15$
$4x^2 - 9x + 26$

$6a^2 -6$
$\ominus 4a^2 \oplus 7a \ominus 3$
$2a^2 + 7a - 9$

Subtract:

1. $6x - 9$
   $3x + 2$

2. $5a + 11$
   $2a + 3$

3. $6y + 4$
   $3y - 10$

4. $17b - 18$
   $13b - 8$

5. $8b - 3$
   $-4b + 7$

6. $8x^2 - y$
   $-5x^2 - 3y$

7. $8b^2 - c^2$
   $6b^2 - 4c^2$

8. $5y^2 - 9y$
   $4y^2 + 8y$

9. $14b^2 - 2c$
   $7b^2 + 9c$

10. $8xy - 2x^2y$
    $-10xy + 4x^2y$

11. $3x^2 - 5x + 4$
    $2x^2 + 7x - 3$

12. $6x^2 + 9x - 1$
    $8x^2 + 6x - 8$

13. $10x^2 + 8x - 3$
    $2x^2 - 3x - 5$

14. $9a^2 - 6a + 7$
    $8a^2 - a + 5$

15. $3y^2 - 7y + 9$
    $5y^2 - 2y + 7$

16. $16a^2 - 7a + 6$
    $8a^2 + 7a - 3$

17. $9a^2 + 4ab - 6b^2$
    $7a^2 - 5ab - 4b^2$

18. $9y^3 - 8y^2 + 7y - 3$
    $3y^3 + 7y^2 + 4y + 8$

19. $6x^2 + 8x + 3$
    $3x^2 -4$

20. $4x^2 - 7x - 3$
    $7x^2 +7$

21. $10a^2 +6$
    $8a^2 - 4a - 3$

22. $3y^2 +6$
    $4y^2 + 6y + 5$

23. $8b^2 - 6b - 4$
    $2b^2 + 7b$

24. $8y^2 + 7y + 7$
    $6y^2 - 4y$

25. $10x^2 - 3x$
    $3x^2 + 9x - 6$

26. $2x^2 + 9x$
    $8x^2 - 5x - 4$

27. $4x^2 - 6x + 5$
    $ 8x - 2$

28. $13x^2 + 7x + 3$
    $ -4x + 7$

Chapter 2 Review - Simplify by addition or subtraction

1. $(3x - 7) + (2x + 9)$

2. $(6x + 10) + (5x - 3) + (x - 1)$

3. $8x^2 - 3x + 14 - 3x^2 + 5x - 7$

4. $11x^2 + 8 + 4x - 1 + x^2 + 16$

5. $(2x - 9) - (x + 3)$

6. $(10x^2 + 5x - 7) - (4x^2 - 9x + 6)$

7. Add: $6x - 8$
          $\underline{3x + 12}$

8. Add: $2x + 9$
          $8x - 7$
          $\underline{-5x + 3}$

9. Subtract: $32x + 26$
              $\underline{10x - 18}$

10. Subtract: $8x^2 - 9x + 20$
               $\underline{7x^2 - 7x - 13}$

11. $x^3 + 3x^2 - 4x + 1 + 2x^3 - x^2 + 7x - 3$

12. Subtract: $2a - 5b + 7c - d$
               $\underline{a + 8b - c + 4d}$

13. Add: $8y + 11z$
          $-3y + 2z$
          $\underline{-y - 8z}$

14. $(6x^2 - 4x + 19) + (7x^2 + 13x - 23)$

15. $(9x^2 + 7x + 13) - (-10x^2 + 5x - 10)$

16. Subtract: $3x^3 - x^2 + 2x + 2$
               $\underline{x^3 + x^2 - 3x + 1}$

17. $(4x^2 - 9) - (x^2 + 7x - 3)$

18. $8x^2 - 9x + 6x + 11 - 7x + 2 - x^2$

19. Add: $6x^3 + 8x^2 - 9x + 12$
          $\underline{3x^3 + 4x^2 - 9x - 16}$

20. $(7x - 3) + (3x - 9) + (-8x + 5)$

Cumulative Review 1

Name the underlined item:

1. <u>12ab</u> _____

2. <u>2x - 11 = 27</u> _____

Identify the number:

3. 6 _____

4. $\sqrt{23}$ _____

Name the property:

5. 8(3x - 7) = 8(3x) + 8(-7) _____

   _____

6. $y \times \frac{1}{y} = 1$ _____

7. 4x + 8 = 20
   <u> - 8   - 8</u> _____

8. 25 = x^2
   x^2 = 25 _____

Simplify:

9. -23 - 18

10. 6 + (7x + 11)

11. $3^2 - 5 \times 2 + 5^2$

12. (4)(-5)(-6) + (-2)(3)(+10)

13. 8 + (7 - 4{6 + 1} + 12) - 9

14. Is "7" the solution to the equation $3x^2 - 20x - 7 = 0$?

15. Find R: $R = \frac{2x^2 - x}{3x - 4}$ when x = 1

16. (2)(-3) - (-4)(-1) - (-2)(6)

Chapter 3
Multiplication of Polynomials

Multiplication of Monomials by Monomials - Three steps must be taken: **1)** multiply the signs, **2)** multiply the numerical coefficients, and **3)** add the exponents of the like variables.

Examples

a. $(5x^2)(+7x^3y^2) = 35x^5y^2$ b. $(-4a^4b)(-5a^2b^3)(-2b^5) = -40a^6b^9$

Multiply the monomials:

1. $y^3 \bullet y^5$

2. $b^6 \bullet b^7$

3. $y^4 \bullet y^{10}$

4. $3x^2(5x^3)$

5. $(2y^5)(6y^2)$

6. $(-3a^2)(4a^6)$

7. $(-5y^4)(-6y^3)$

8. $+9a^5(2a^7)$

9. $(-2b^2c^3)(-4b^3c)$

10. $(-3c)(+5c)$

11. $(+7x^4)(6x^5)$

12. $(10x^3y)(-4x^2y^2)$

13. $(-9x)(7x^6)$

14. $(-6x^4)(\frac{1}{2}x^3)$

15. $(+5y^3z^2)(2yz^6)$

16. $(-8y^7a)(-4yz^2a^2)$

17. $(-4x^3)(2x^2)(5x^4)$

18. $(3x^5)(-3x)(-10x^7)$

19. $(+10b^2)(3b^6)(-2b)$

20. $(-12b^5)(-2b^4)(-\frac{1}{2}b^2)$

21. $b^2c^4 \bullet b^5c^7$

22. $(x^3y^5)(-xy^7)$

23. $(4a^4)(-2a)(\frac{1}{4}a^3)$

24. $(-10x^2)(-5x^3)(\frac{1}{2}xy)$

25. $(3ab)(-4bc)(2ac)$

26. $(4xy)(2yz)(6xz)$

27. $(-2ab^2)(-3bc^2)(-4ca^2)$

28. $(9x)(-7xy^2z)$

29. $(-8x^{10}y)(-5x^8y^{10})$

30. $(4y^5)(+3y^4z)$

31. $(10a^7)(-6abc)(-2b^2c^4)$

32. $(-5z^4)(-4z^4)(-3z)$

33. $(r^3)(3rst^2)(-2st^4)$

31

Multiplication of Monomials with Exponents outside of Parentheses - The exponent outside a parentheses tells you how many times to write down the parentheses and multiply.

Examples

a. $(4a^4)^3$
    $= (4a^4)(4a^4)(4a^4)$
    $= 64a^{12}$

b. When there is no numerical coefficient, multiply the exponents inside the parenthese by the exponent outside the parenthese.
   $(x^5 y^7)^8 = x^{40} y^{56}$

Multiply:

1. $(2r^3b)^2$
2. $(6x^2y)^3$
3. $(-2s^2t^3)^4$
4. $(3xy^5)^4$
5. $(5ab^2)^3$
6. $(-3x^2y^2)^3$
7. $(3x^2)^4$
8. $(5a^3)^3$
9. $(-2x^3y)^2$

10. $(-3a^2b^3)^3$
11. $(3ab)^2$
12. $(4x^2y^2)^4$
13. $(2xy^4)^3$
14. $(-5x^2y)^2$
15. $(2a^6)^4$
16. $(a^5)^2$
17. $(x^4)^3$
18. $(x^3)^4$

19. $(a^2b^3)^3$
20. $(x^4y)^5$
21. $(-x^2y^4)^4$
22. $(-a^6b^7)^6$
23. $(-a^3)^5$
24. $(-b^2c^5)^5$
25. $(xyz^2)^4$
26. $(x^8y^4)^6$
27. $(-a^5b^6)^4$

28. $(-a^8c^7)^7$
29. $(x^4z^2)^8$
30. $(z^3)^{10}$
31. $(6x^2)^3$
32. $(x^5)^7$
33. $(4a^2b)^3$
34. $(xy^3)^8$
35. $(-3b^2c^2)^2$
36. $(-x^3y^4)^5$

Multiplication of Monomials and Monomials with Exponents outside the Parentheses - These problems are more complicated but no new skills are required to work them.

Examples

a. $(3a^2b^2)^3(-2ab)^2$
 $= (3a^2b^2)(3a^2b^2)(3a^2b^2)(-2ab)(-2ab)$
 $= 108a^8b^8$

b. $(-2x^4)^3(x^2y^5)^6$
 $= (-2x^4)(-2x^4)(-2x^4)(x^{12}y^{30})$
 $= -8x^{24}y^{30}$

Multiply:

1. $(2ab)^2(a^3)^3$

2. $(3xy^2)^2(x^2)^3$

3. $(-4x^2)^3(b)^4$

4. $(-2xy)^2(xy^2)^2$

5. $(6y^2)^2(-3y)^3$

6. $(4a)^3(-2x)^2$

7. $(4ab^4)^3(2x^2y^3)^2$

8. $(3x^2y)^4(4x^2y)$

9. $(-6x)^2(x^3y^4)^5$

10. $(a^2b^3)^4(-3a)^2$

11. $(a^5)^4(-2a^2)^3$

12. $(xy^3)^3(x^2y^4)^5$

13. $(a^3b^2)^4(-ab^3)^3$

14. $(6xy^3)^2(-x^2y^2)$

15. $(3ab^4)^2(-a^3b)$

16. $(4x^2y)(-3xy^2)^3$

17. $(3x^3y)^4(3xy)$

18. $(3a^2b^2)^3(2ab)^2$

19. $(-2y)^3(x^2y^6)^4$

20. $(-6x^2)^3(x^5)^7$

21. $(2ab)(-3a^4b)^3$

22. $(3xb)^2(-2x^3b^2)^3$

23. $(-4y^3)^2(x^4y^5)^4$

24. $(a^2b^4)^5(b^3c^6)^4$

Multiplication of Monomials and Polynomials - When written horizontally, each term inside the parentheses must be multiplied by the term outside the parentheses. (Previously learned as the distributive property.) In the examples below, the second step can be left out when working the problem.

Examples

a. $4x(2x - 5)$
    $= 4x(2x) + 4x(-5)$
    $= 8x^2 - 20x$

b. $3a^2(4a + 2ab - 6b)$
    $= 3a^2(4a) + 3a^2(2ab) + 3a^2(-6b)$
    $= 12a^3 + 6a^3b - 18a^2b$

---

1. $2x(x - 3)$

2. $3(2a + 5)$

3. $3y(y + 7)$

4. $6(x - 7)$

5. $4b(2b + 7)$

6. $3y(5y - 7)$

7. $6x(2x^2 - 3x)$

8. $4a(-6a^2 + 2)$

9. $xy(x + y)$

10. $2ab(3a - 5b)$

11. $-4(2x - 3)$

12. $-6a(5a + 7)$

13. $-2xy(3x - 5y)$

14. $-10ab^2(2 + 5ab)$

15. $-6x^3y^2(6x - 3y)$

16. $-5a^3(3a - 5b)$

17. $-7x^4(2a - 3y)$

18. $3(2x^2 - 5x + 3)$

19. $4a(6a^2 - 4a - 3)$

20. $2x(3x + 2y - 5z)$

21. $-5y(y^2 + 7y - 6)$

22. $4x^2(x^2 - 6x + 9)$

23. $3a^2(2a^2 - 7a - 6)$

24. $-4y^2(3y^2 + 9y + 1)$

Multiplication of Monomials and Polynomials - These problems can also be written vertically. Multiply each term on the top by the term on the bottom. The bottom term may be on the left side or the right side.

Examples

a. $\quad 3x^2 + 5x - 9$
$\quad\quad\quad\underline{\quad\quad\quad 4x}$
$\quad 12x^3 + 20x^2 - 36x$

b. $\quad 3a^2 - 14ab + 10b^2$
$\quad\quad\underline{\quad -6ab\quad\quad\quad\quad}$
$\quad -18a^3b + 84a^2b^2 - 60ab^3$

Multiply:

1. $3a - 5$
   $\underline{\quad 3\quad}$

2. $6x + 2$
   $\underline{\quad -7\quad}$

3. $4b + 7$
   $\underline{\quad 2b\quad}$

4. $5x^2 - 7x + 8$
   $\underline{\quad\quad\quad 7}$

5. $4x^2 + 11x - 4$
   $\underline{\quad\quad\quad -8}$

6. $8x^2 + 7xy - y^2$
   $\underline{\quad\quad\quad xy}$

7. $8y^2 - 5y + 9$
   $\underline{\quad\quad\quad 6y}$

8. $4a - 9 - 2b$
   $\underline{\quad\quad 3ab}$

9. $9a + 5b + 11c$
   $\underline{\quad\quad\quad -ac}$

10. $5a^2 + 8a - 1$
    $\underline{\quad\quad\quad 9a^2}$

11. $6a - 4b + 7c$
    $\underline{\quad\quad 5bc}$

12. $12a^2 - b^2$
    $\underline{\quad -4ab}$

13. $2x^2y^2 - 8xy$
    $\underline{\quad\quad 2xy}$

14. $7b + 16y$
    $\underline{\quad -by}$

15. $8x^2 + 9x + 20$
    $\underline{\quad\quad\quad -7x}$

16. $3y - 1$
    $\underline{\quad 4\quad}$

17. $6x + 12$
    $\underline{\quad -5\quad}$

18. $9a - 4x$
    $\underline{\quad 3x\quad}$

19. $7x^2 - 4x + 1$
    $\underline{\quad 3x\quad\quad\quad}$

20. $9y^2 + 9y - 7$
    $\underline{\quad -2y\quad\quad}$

21. $2x^2 + 3xy + 8y^2$
    $\underline{\quad xy\quad\quad\quad}$

22. $x^2 + 4xy - 8y^2$
    $\underline{\quad -5xy\quad\quad}$

23. $2a - 6b + 7c$
    $\underline{\quad -abc\quad\quad}$

24. $9xy + 2yz - xz$
    $\underline{\quad 7xz\quad\quad}$

Multiplication of Binomials and Binomials - The <u>FOIL</u> method is used when the problem is written <u>horizontally</u>. Each of the four letters stands for one step in the multiplication process. The **F** tells you to multiply the First terms of each binomial. The **O** tells you to multiply the Outside terms of each binomial. The **I** tells you to multiply the Inside terms of each binomial, and the **L** tells you to multiply the Last terms of each binomial. Normally the **OI** products will be <u>like terms</u> and will be combined to get the final answer.

Examples

a. $(x + 8)(x - 3)$
    F O I L
$= x^2 - 3x + 8x - 24$
$= x^2 + 5x - 24$

b. $(5x - 7)(2x - 9)$
    F O I L
$= 10x^2 - 45x - 14x + 63$
$= 10x^2 - 59x + 63$

Multiply using the FOIL method:

1. $(x + 2)(x + 4)$

2. $(x + 5)(x - 7)$

3. $(x - 3)(x - 6)$

4. $(x - 7)(x + 1)$

5. $(x + 1)(x - 2)$

6. $(x + 3)(x + 4)$

7. $(x - 2)(x - 6)$

8. $(x - 5)(x + 7)$

9. $(x + 9)(x - 3)$

10. $(x + 10)(x + 5)$

11. $(x - 7)(x + 9)$

12. $(2x - 1)(x - 4)$

13. $(3x + 2)(x + 6)$

14. $(5x + 4)(x - 1)$

15. $(2x - 3)(x + 7)$

16. $(4x - 5)(x - 2)$

17. $(6x + 5)(x + 3)$

18. $(5x - 1)(x + 1)$

19. $(x - 2)(3x + 2)$

20. $(x + 8)(2x - 3)$

21. $(x + 4)(4x - 1)$

22. $(x - 3)(2x + 5)$

23. $(x - 8)(4x - 3)$

24. $(x + 6)(2x + 1)$

Multiplication of binomials and binomials continued

25. $(x - 9)(3x - 5)$

26. $(4x - 1)(2x + 3)$

27. $(2x - 5)(7x - 1)$

28. $(6x + 1)(6x + 5)$

29. $(8x + 3)(3x - 8)$

30. $(7x - 8)(4x + 3)$

31. $(3x + 5)(8x - 7)$

32. $(5x - 4)(5x - 6)$

33. $(x - 9)(3x + 4)$

34. $(x - 3)(x + 10)$

35. $(2x - 9)(x + 2)$

36. $(x + 10)(2x - 5)$

37. $(2x + 1)(3x - 1)$

38. $(3x - 7)(x + 1)$

39. $(x - 6)(x + 5)$

40. $(x + 9)(x + 4)$

41. $(7x - 2)(x + 1)$

42. $(x - 4)(7x - 6)$

43. $(3x - 1)(x + 7)$

44. $(6x + 5)(4x - 3)$

45. $(4x + 1)(3x - 1)$

46. $(8x - 3)(x + 2)$

47. $(3x + 1)(x - 7)$

48. $(4x - 3)(3x - 2)$

49. $(x - 1)(x + 1)$

50. $(x + 6)(x - 6)$

51. $(2x - 5)(2x + 5)$

52. $(6x + 1)(6x - 1)$

52. $(3x + 4)(3x + 4)$

53. $(5x - 2)(5x - 2)$

54. $(7a + 3)(7a + 3)$

55. $(x - a)(x + a)$

56. $(a + b)(a + b)$

57. $(x - y)(x - y)$

Multiplication of Binomials and Binomials written vertically - When written vertically, each term on the top is multiplied by each term on the bottom. Begin with either the right bottom term or the left bottom term.

Examples

Starting with the
right bottom term.(-4)

$$2x + 3$$
$$\underline{x - 4}$$
$$-8x - 12$$
$$\underline{2x^2 + 3x}$$
$$2x^2 - 5x - 12$$

Starting with the
left bottom term.(3x)

$$6x - 1$$
$$\underline{3x - 5}$$
$$18x^2 - 3x$$
$$\underline{-30x + 5}$$
$$18x^2 - 33x + 5$$

Multiply:

1. x + 6
   <u>x + 4</u>

2. x + 9
   <u>x - 5</u>

3. x - 4
   <u>x + 2</u>

4. x - 6
   <u>x - 5</u>

5. x + 7
   <u>x - 4</u>

6. 2x + 8
   <u>x + 7</u>

7. 3x - 1
   <u>x + 6</u>

8. 4x - 3
   <u>x - 1</u>

9. 6x + 2
   <u>x - 8</u>

10. 5x - 3
    <u>x + 3</u>

11. x - 2
    <u>2x + 3</u>

12. x + 4
    <u>3x + 5</u>

13. x + 10
    <u>4x - 1</u>

14. x - 6
    <u>5x - 7</u>

15. x + 1
    <u>2x - 3</u>

16. 4x - 1
    <u>2x + 7</u>

17. 7x + 8
    <u>4x + 9</u>

18. 2x - 6
    <u>5x - 6</u>

19. 5x + 3
    <u>2x - 8</u>

20. 3x + 7
    <u>4x - 5</u>

Multiplication of Binomials and Trinomials - In these problems, every term of the top is multiplied by every term on the bottom. Begin with the <u>right bottom term</u> **or** <u>the left bottom term</u>.

Examples

a. Begin with the left bottom term. →
$$2x^2 + 9x - 1$$
$$\underline{3x - 5}$$
$$6x^3 + 27x^2 - 3x$$
$$\underline{-10x^2 - 45x + 5}$$
$$6x^3 + 17x^2 - 48x + 5$$

$$4x^2 - 7x + 2$$
$$\underline{2x - 3} ←$$
$$-12x^2 + 21x - 6$$
$$\underline{8x^3 - 14x^2 + 4x}$$
$$8x^3 - 26x^2 + 25x - 6$$

Begin with the right bottom term.

Multiply:

1. $x^2 + 5x - 6$
   $\underline{x - 3}$

2. $x^2 + 7x + 3$
   $\underline{x + 7}$

3. $x^2 - 4x - 7$
   $\underline{x + 4}$

4. $3x^2 - 9x + 2$
   $\underline{x - 2}$

5. $2x^2 - 8x + 4$
   $\underline{x + 6}$

6. $5x^2 + 7x + 11$
   $\underline{x + 9}$

7. $3x^2 - 5x - 6$
   $\underline{8x - 1}$

8. $8x^2 + 7x - 1$
   $\underline{4x - 8}$

9. $4x^2 + 5x - 1$
   $\underline{3x + 1}$

10. $3x^2 + 2x + 8$
    $\underline{7x - 3}$

11. $7x^2 - 8x - 4$
    $\underline{x + 7}$

12. $6x^2 - 4x + 9$
    $\underline{x - 3}$

13. $2x^2 - 8x - 8$
    $\underline{x + 5}$

14. $x^2 + 4x - 1$
    $\underline{x - 2}$

15. $x^2 + 9x + 2$
    $\underline{9x + 6}$

16. $7x^2 - 5x + 3$
    $\underline{4x - 5}$

17. $5x^2 + 4x - 3$
    $\underline{2x + 7}$

18. $5x^2 - 7x - 10$
    $\underline{3x - 4}$

19. $9x^2 + 2x + 1$
    $\underline{x - 8}$

20. $4x^2 - 7x - 2$
    $\underline{3x + 2}$

Chapter 3 Review

1. $6x(4x-7)$

2. $(x-5)(2x+9)$

3. $(3xy^2)^3$

4. $2a^2+8a-12$
   $\underline{3a+2}$

5. $(4x-5)(3x-2)$

6. $2x^2-x-11$
   $\underline{5x}$

7. $(4a^3b^2)(-7ab^6)$

8. $x-9$
   $\underline{x+3}$

9. $4x^2+5x-1$
   $\underline{6x+7}$

10. $(a^3b^5)^8$

11. $2x^2y(6x^2+4xy-8y^2)$

12. $(3a^2b)^3(a^4b^5)^6$

13. $4x+3$
    $\underline{2x+9}$

14. $(5x^4y^3)(-3x^2y^6)$

15. $8a^3-a^2+7a+16$
    $\underline{-3a}$

16. $(x-9)(x-10)$

40

Cumulative Review 2

Add or Subtract:

1. $(3x + 4) + (7x - 11)$

2. $(6x^2 - 4x + 10) + (2x^2 - 9x - 3)$

3. $8x^2 + 9x - 2 - 7x + 10 + x - x^2$

4. $(3x - 2) - (x - 7)$

5. $(5x^2 + 6x - 1) - (2x^2 + 3x + 13)$

6. $4x^2 - 16x - 7$
   $\underline{3x^2 + 11x - 3}$

7. Subtract: $7x^2 + 9x + 6$
   $\underline{x^2 + 4x - 9}$

8. $6x + 12$
   $-3x + 2$
   $\underline{7x - 9}$

9. $(4x^2 - 21) - (7x - 15)$

Name the underlined:

10. 5x<u>3</u> _____

11. 7<u>a</u> _____

Identify the number:

12. - 4 _____

13. .75 _____

Name the property:

14. $4x + (-4x) = 0$ _____

15. $3 \times 2 \times 7 = 2 \times 7 \times 3$ _____
_____

Simplify:

16. $19 - (5x - 11)$

17. $10 - 3^2 + 12 \div 2^2 - 4$

18. $6 + (12 + \{8 - 14\} - 9) - 17$

19. Is "$\frac{1}{2}$" the solution to the equation
    $4x - 7 = 2x - 6$?

Chapter 4
Division of Polynomials & Negative Exponents

Zero exponents - any number, variable, or entire term raised to the zero power is equal to "1". The only exception to this rule is "0" to the "0" power.

Examples

a. $x^0 = 1$ b. $x^0 y = 1y = y$ c. $\frac{(ab)^0}{b} = \frac{1}{b}$ d. $a^3 b^0 c = a^3 c$

---

1. $a^0 =$
2. $y^0 =$
3. $r^0 =$
4. $(xz)^0 =$
5. $(ax)^0 =$
6. $(xyz)^0 =$

7. $x^0 b =$
8. $r^0 c =$
9. $xc^0 =$
10. $a^2 b^0 =$
11. $\frac{a^0}{c} =$
12. $\frac{x^0 b}{5}$

13. $a^2 b^0 x^2 =$
14. $3xy^0 =$
15. $-xy^0 =$
16. $(3a^2)^0 =$
17. $3(ab)^0 =$
18. $-b(c^2 d)^0 =$

==========

Dividing Monomials into Monomials - three steps must be taken: **1)** divide the signs, **2)** divide the numerical coefficients, and **3)** <u>subtract</u> the exponents of the like variables.

Examples

a. $\frac{-8a^5}{-2a^2} = 4a^3$ b. $15y^7 \div (-3y^3) = -5y^4$ c. $2a^2 b^3 \overline{)-8a^9 b^7} \quad -4a^7 b^4$

Divide

1. $\frac{x^6}{x^2}$
2. $\frac{a^{10}}{a^4}$
3. $\frac{b^5}{-b^4}$

4. $x^9 \div (-x^5)$
5. $(a^4 b^3) \div (-a^2 b^2)$
6. $y^7 z^3 \div y^2 z$

7. $-x^3 \overline{)x^6}$
8. $a^2 c^3 \overline{)a^4 c^7}$
9. $-b^4 x^5 \overline{)-b^6 x^9}$

Division of Monomials Continued

10. $\dfrac{-a^8}{a^5}$

11. $\dfrac{-9a^5}{-a^2}$

12. $\dfrac{x^{10}y^7}{-x^9 y}$

13. $\dfrac{10a^4}{-5a}$

14. $\dfrac{-12x^4}{-3x^3}$

15. $\dfrac{8b^7 c^2}{2b^6 c^2}$

16. $\dfrac{20x^4 y^8}{-5x^2}$

17. $\dfrac{9a^3 b^2}{3a^3}$

18. $\dfrac{-18r^5 t^6}{2rt^3}$

19. $\dfrac{-30x^6 t^{10}}{10x^4 t^9}$

20. $\dfrac{25a^4 b^9}{5a^2 b^3}$

21. $(-12x^7) \div 3x^2$

22. $y^{12} \div y^5$

23. $-a^6 c^4 \div a^2 c^3$

24. $24x^5 \div (-12x)$

25. $28a^3 b \div 7ab$

26. $(-10x^4 b^6) \div (-5x^3 b^3)$

27. $22b^6 x^4 \div (-11b^5 x^4)$

28. $14ac^4 \div 2c^4$

29. $24x^4 c^2 \div (-3x^4)$

30. $a^7 \div a^3$

31. $15a^3 c^8 \div 3a^3 c^7$

32. $12a^5 \overline{)24a^6}$

33. $9x \overline{)36x^8}$

34. $-x^4 y^2 \overline{)x^7 y^2}$

35. $-2b^7 c \overline{)8b^9 c^5}$

36. $6b^2 y^2 \overline{)24b^6 y^2}$

37. $2c^6 \overline{)-16a^4 c^7}$

38. $-5z^4 \overline{)-25z^4 b}$

39. $6x^8 \overline{)30x^9 c}$

40. $-9a^4 \overline{)-18a^8 x^4}$

41. $8b^2 c^2 x^2 \overline{)32b^4 c^2 x^7}$

42. $-4x^6 \overline{)12b^9 x^8}$

More Division of Monomials - when a denominator cannot divide <u>evenly</u> into the numerator, the coefficients might possibly be reduced, and the exponents of the variables can be subtracted. (If the largest exponent is a numerator, the variable stays a numerator, but if the largest exponent is a denominator, then the variable stays a denominator.)

Examples

a. $\dfrac{a^5}{a^7} = \dfrac{1}{a^2}$ b. $\dfrac{-x^4 c^5}{-x^6 c^4} = \dfrac{c}{x^2}$ c. $\dfrac{6a^2}{-8} = -\dfrac{3a^2}{4}$ d. $\dfrac{9a^4 b^3}{6ab^5} = \dfrac{3a^3}{2b^2}$

Simplify

1. $\dfrac{a^3}{a^6}$

2. $\dfrac{b^6}{b^7}$

3. $\dfrac{z^3}{-z^8}$

4. $\dfrac{-c}{-c^4}$

5. $\dfrac{-z^3}{z^5}$

6. $\dfrac{-w^2}{-w^8}$

7. $\dfrac{x^6 c^5}{x^3 c^6}$

8. $\dfrac{r^8 t^4}{r^2 t^7}$

9. $\dfrac{xz^4}{-x^4 z}$

10. $\dfrac{c^4 r^3}{c^2 r^8}$

11. $\dfrac{2a^4}{6}$

12. $\dfrac{6y^3}{9}$

13. $\dfrac{10y^2}{15b^2}$

14. $\dfrac{12z^3}{9c^4}$

15. $\dfrac{6x^5}{15x^6}$

16. $\dfrac{3a^4 b^3}{-9a^3 b^4}$

17. $\dfrac{9x^3 z^5}{-15x^7 z^4}$

18. $\dfrac{6x^5 y^4}{9x^6 y^3}$

19. $\dfrac{-b^5}{-b^7}$

20. $\dfrac{4a^6 b^2}{6ab^4}$

21. $\dfrac{6x^2}{21}$

22. $\dfrac{-8b^4 c^2}{6bc^8}$

23. $\dfrac{-a^3 b^4}{-ab^7}$

24. $\dfrac{c^4}{-c^7}$

Dividing Monomials into Polynomials - when a monomial is divided into a polynomial, the monomial is divided into each term of the polynomial.

Example

a. $\dfrac{3x^2-9x}{3x} = x - 3$ b. $\dfrac{-5b^2+20b^3+35b^4}{-5b} = b - 4b^2 - 7b^3$

Divide

1. $\dfrac{3a+12}{3}$

2. $\dfrac{5x^2-4x}{x}$

3. $\dfrac{15a^3-3a^2}{3a}$

4. $\dfrac{5x^3-15x}{5x}$

5. $\dfrac{16a^2-2ab}{2a}$

6. $\dfrac{30z^2-15za+20a^2}{5}$

7. $\dfrac{25a^3b^3-5a^2b^2+10ab}{5ab}$

8. $\dfrac{6y^2-5y}{y}$

9. $\dfrac{18x^2-6x}{6x}$

10. $\dfrac{8a^3-40a}{8a}$

11. $\dfrac{12x^2-24xz}{4x}$

12. $\dfrac{7x-35}{7}$

13. $\dfrac{12x^2-6x}{3x}$

14. $\dfrac{21b^3+28b^2-7b}{-7b}$

15. $\dfrac{16x^4z^5-32x^5z^4}{8x^3z^3}$

16. $\dfrac{x^4y^3+x^3y^4}{-x^3y^3}$

17. $\dfrac{a^6b^5-a^5b^4+a^4b^3}{a^4b^3}$

18. $\dfrac{21a^4b^3-18a^3b^4}{-3a^2b}$

Dividing Binomials into Trinomials - this involves many steps. Study the instructions and the example.

Example

1) Divide the first term of the divisor into first term of the dividend.
2) Put the quotient above the dividend.
3) Multiply the quotient and the divisor, and put the product under the dividend.
4) Subtract (change signs) then bring down the next term.
5) Go back to step 1 and repeat the process.

$$x-2 \overline{\smash{)}\, x^2 + 9x - 22}$$

Quotient: $x + 11$

Working:
$\underline{\pm x^2 \mp 2x}$
$\qquad 11x - 22$
$\qquad \underline{\pm 11x \mp 22}$

Divide

1. $x+2 \,\overline{\smash{)}\, x^2 + 8x + 12}$

2. $x-3 \,\overline{\smash{)}\, x^2 - 7x + 12}$

3. $x+6 \,\overline{\smash{)}\, x^2 + 4x - 12}$

4. $x-3 \,\overline{\smash{)}\, x^2 + 3x - 18}$

5. $x+1 \,\overline{\smash{)}\, x^2 - 4x - 5}$

6. $x-7 \,\overline{\smash{)}\, x^2 - 14x + 49}$

7. $x-2 \,\overline{\smash{)}\, x^2 + 7x - 18}$

8. $x-5 \,\overline{\smash{)}\, x^2 + x - 30}$

More Dividing Binomials into Trinomials - these problems will have a remainder.

1. $x-3 \overline{\smash{\big)}\ x^2 + 4x - 26}$

2. $x+7 \overline{\smash{\big)}\ x^2 + x - 39}$

3. $x-6 \overline{\smash{\big)}\ x^2 - 5x - 15}$

4. $x+10 \overline{\smash{\big)}\ x^2 + 19x + 80}$

5. $x-8 \overline{\smash{\big)}\ x^2 - 15x - 41}$

6. $x-7 \overline{\smash{\big)}\ x^2 - 3x - 32}$

7. $x+9 \overline{\smash{\big)}\ x^2 - 3x - 20}$

8. $x-5 \overline{\smash{\big)}\ x^2 - 3x + 2}$

9. $x-10 \overline{\smash{\big)}\ x^2 - 15x + 65}$

10. $x+8 \overline{\smash{\big)}\ x^2 + 17x + 70}$

More Dividing Binomials into Binomials - in these problems, the first term of the divisor and/or dividend will have a numerical coefficient.

Examples

a.
$$\begin{array}{r} 2x + 11 \\ x-4 \overline{\smash{\big)}\, 2x^2 + 3x - 44} \\ \underline{\ominus 2x^2 \ominus 8x} \\ +11x - 44 \\ \underline{\oplus 11x \ominus 44} \end{array}$$

b.
$$\begin{array}{r} x + 4 \\ 2x+1 \overline{\smash{\big)}\, 2x^2 + 9x + 4} \\ \underline{\ominus 2x^2 \ominus x} \\ 8x + 4 \\ \underline{\ominus 8x \ominus 4} \end{array}$$

Divide

1. $x+3 \,\overline{\smash{\big)}\, 2x^2 + 9x + 9}$

2. $x-4 \,\overline{\smash{\big)}\, 3x^2 - 11x - 4}$

3. $6x+1 \,\overline{\smash{\big)}\, 6x^2 + 31x + 5}$

4. $3x-10 \,\overline{\smash{\big)}\, 3x^2 - 4x - 20}$

5. $x-4 \,\overline{\smash{\big)}\, 5x^2 - 18x - 8}$

6. $3x+5 \,\overline{\smash{\big)}\, 3x^2 + 20x + 25}$

7. $x+9 \,\overline{\smash{\big)}\, 3x^2 + 26x - 9}$

8. $3x-2 \,\overline{\smash{\big)}\, 3x^2 - 11x + 6}$

More Dividing Binomials into Trinomials Continued

9. $x-7 \overline{) 5x^2 - 42x + 49}$

10. $4x+3 \overline{) 4x^2 + 7x + 3}$

11. $x+8 \overline{) 7x^2 + 50x - 58}$

12. $2x+9 \overline{) 2x^2 - 3x - 50}$

13. $x-3 \overline{) 4x^2 - 17x + 15}$

14. $2x-1 \overline{) 2x^2 + 3x - 8}$

15. $x+4 \overline{) 7x^2 + 32x + 10}$

16. $3x+5 \overline{) 3x^2 + 2x - 5}$

17. $4x-1 \overline{) 4x^2 + 23x - 6}$

18. $x-1 \overline{) 5x^2 + 8x + 12}$

Dividing Binomials into Polynomials - example (b) shows how to divide a binomial into a 4-term polynomial. The same procedures are followed, but these problems go one more step.

Examples

a. $2x+1 \overline{\smash{\big)}\ 16x^2 + 14x + 3}$ with quotient $8x + 3$

b. $x+3 \overline{\smash{\big)}\ x^3 + 7x^2 + 8x - 12}$ with quotient $x^2 + 4x - 4$

Divide

1. $3x+5 \overline{\smash{\big)}\ 21x^2 + 23x - 20}$

2. $5x-2 \overline{\smash{\big)}\ 15x^2 - x - 2}$

3. $x-6 \overline{\smash{\big)}\ x^3 - 4x^2 - 5x - 42}$

4. $x+2 \overline{\smash{\big)}\ x^3 - x^2 - 4x + 4}$

5. $6x-7 \overline{\smash{\big)}\ 36x^2 - 36x - 7}$

6. $x+4 \overline{\smash{\big)}\ 4x^3 + 15x^2 + 3x + 28}$

Dividing Binomials into Polynomials Continued

7. $3x+4 \overline{)\ 6x^2 - 7x - 20\ }$

8. $x-6 \overline{)\ 2x^3 - 14x^2 + 6x + 36\ }$

9. $x+9 \overline{)\ 2x^3 + 14x^2 - 39x - 27\ }$

10. $7x-1 \overline{)\ 56x^2 - 29x + 10\ }$

11. $x-7 \overline{)\ 4x^3 - 29x^2 + 14x - 49\ }$

12. $2x+5 \overline{)\ 10x^2 + 29x + 10\ }$

13. $x+2 \overline{)\ 3x^3 + 11x^2 + 8x - 4\ }$

14. $7x-2 \overline{)\ 21x^2 + 36x - 12\ }$

Division with Missing Exponents - in some problems, the dividend may have a missing exponent. For example, in the polynomial $4x^3 + 2x - 1$, there is no x^2. To divide a polynomial containing a missing exponent, insert the variable with the missing exponent into the polynomial and <u>give it a numerical coefficient of "0"</u>, (insert $0x^2$ between $4x^3$ and $2x$), then perform the division.

Example

$$\frac{3x^3-4x-16}{x-2} = \frac{3x^3+0x^2-4x-16}{x-2} \longrightarrow$$

$$\begin{array}{r} 3x^2 + 6x + 8 \\ x-2 \overline{\smash{\big)}\, 3x^3 + 0x^2 - 4x - 16} \\ \underline{3x^3 - 6x^2} \\ 6x^2 - 4x \\ \underline{6x^2 - 12x} \\ 8x - 16 \\ \underline{8x - 16} \end{array}$$

Divide

1. $\dfrac{2x^3-53x+15}{x-5}$

2. $\dfrac{16x^3-2x-1}{2x-1}$

3. $\dfrac{2x^3+19x^2-36}{2x+3}$

4. $\dfrac{4x^3-x^2-28}{x-2}$

5. $\dfrac{2x^3-37x+20}{x-4}$

6. $\dfrac{x^3-5x+100}{x+5}$

Negative Exponents - in algebra there are <u>negative exponents</u> as well as positive exponents. To change a negative exponent to a positive exponent, move the exponent and its base across the fraction bar and remove the negative sign.

Examples

a. $3^{-2} = \dfrac{1}{3^2} = \dfrac{1}{9}$ b. $\dfrac{2}{x^{-3}} = \dfrac{2x^3}{1} = 2x^3$ c. $\dfrac{x^5}{x^{-3}} = \dfrac{x^5 x^3}{1} = x^8$

Change all negative exponents to positive exponents and simplify

1. 4^{-3}

2. 6^{-2}

3. 10^{-1}

4. x^{-5}

5. $\dfrac{6}{c^{-2}}$

6. $\dfrac{10}{x^{-3}}$

7. $\dfrac{8}{b^{-2}}$

8. $\dfrac{b^5}{b^{-2}}$

9. $\dfrac{a^4}{a^{-4}}$

10. $\dfrac{2x^4}{x^{-2}}$

11. $\dfrac{4c^6}{c^{-3}}$

12. $\dfrac{x^2 y^2}{x^{-3} y^{-5}}$

13. $\dfrac{a^3 b^4}{a^{-1} b^{-4}}$

14. $\dfrac{x^5 y^3}{x^{-4} y^{-2}}$

15. $\dfrac{r^3 t}{r^{-4} t^{-1}}$

16. $\dfrac{a^6}{a^{-1}}$

17. $\dfrac{3x^2 y^4}{x^{-4} y^{-2}}$

18. $\dfrac{a^4}{a^{-2} b^{-5}}$

19. $\dfrac{b^4}{c^{-3}}$

20. $a^{-4} b^4$

21. $\dfrac{x^2}{x^{-9} y^{-1}}$

22. $\dfrac{3a^2}{b^{-4}}$

23. $\dfrac{ab}{a^{-2} b^{-5}}$

24. $x^{-1} y^{-5}$

25. $\dfrac{a^2 b^7}{a^{-2}}$

26. $c^{-2} z^{-3}$

27. $\dfrac{b^4 c^3}{c^{-2}}$

28. $\dfrac{9x}{y^{-6}}$

More Negative Exponents - remember the basic rule for negative exponents: <u>every term with a negative exponent must cross the fraction bar.</u>

Examples

a. $\dfrac{a^{-2}}{b^{-3}} = \dfrac{b^3}{a^2}$
b. $\dfrac{3^{-2}x^5}{2x^{-3}} = \dfrac{x^5 x^3}{3^2(2)} = \dfrac{x^8}{9(2)} = \dfrac{x^8}{18}$
c. $\dfrac{a^3 c^{-4}}{a^{-2} c} = \dfrac{a^3 a^2}{c^4 c} = \dfrac{a^5}{c^5}$

Change all positive exponents to negative exponents and simplify

1. $\dfrac{a^{-2}}{y^{-2}}$

2. $\dfrac{y^{-2}}{a^{-1}}$

3. $\dfrac{c^{-5}}{r^{-2}}$

4. $\dfrac{4^{-1} b^2}{3 b^{-3}}$

5. $\dfrac{5^{-1} r^6}{2^2 r^{-3}}$

6. $\dfrac{2^{-2} x^2}{3^2 x^{-2}}$

7. $\dfrac{a^4 b^{-3}}{a^{-1} b^4}$

8. $\dfrac{x^5 y^{-2}}{x^{-3} y^6}$

9. $\dfrac{a^4 c^{-2}}{a^{-1} c^3}$

10. $\dfrac{ab^{-3}}{a^{-2} b^2}$

11. $\dfrac{4 c^3}{3^{-1} c^{-1}}$

12. $\dfrac{x^5 y^2 c^{-3}}{x^{-1} c^2}$

13. $\dfrac{a^{-5}}{3^{-2}}$

14. $\dfrac{a^{-3} b^3 x^4}{a^2 b^3}$

15. $\dfrac{7^{-1} x^4}{3^{-2} x^{-2}}$

16. $\dfrac{4^{-2}}{c^{-3}}$

17. $\dfrac{x^{-1} y^{-4} z^3}{x^2 y^3 z^{-2}}$

18. $\dfrac{a^4 b c^2}{a^{-2} b^{-3} c^2}$

19. $\dfrac{5^{-2}}{r^{-4}}$

20. $\dfrac{5 b^5}{2^{-1} b^{-4}}$

21. $\dfrac{x^{-6} y}{x^4 y^{-4}}$

Negative Exponents, with Exponents Outside Parentheses - to simplify, follow these steps: **1)** multiply each exponent inside the parentheses by the exponent outside the parentheses, **2)** change all negative exponents to positive exponents by crossing the fraction bar, **3)** simplify by dividing and/or multiplying the like variables.

Examples

a. $\left(\dfrac{x^2}{x^3}\right)^4 = \dfrac{x^8}{x^{12}} = \dfrac{1}{x^4}$

b. $\left(\dfrac{2}{x^8}\right)^{-2} = \dfrac{2^{-2}}{x^{-16}} = \dfrac{x^{16}}{2^2} = \dfrac{x^{16}}{4}$

c. $\left(\dfrac{a^{-2}b}{a^2 b^3}\right)^{-3} = \dfrac{a^6 b^{-3}}{a^{-6} b^{-9}} = \dfrac{a^6 a^6 b^9}{b^3} = a^{12} b^6$

Remove the parentheses and simplify

1. $\left(\dfrac{a^2}{a^3}\right)^2$

2. $\left(\dfrac{x^4}{x^5}\right)^4$

3. $\left(\dfrac{b^3}{b^2}\right)^{-3}$

4. $\left(\dfrac{3}{a^2}\right)^{-2}$

5. $\left(\dfrac{4}{x^3}\right)^{-2}$

6. $\left(\dfrac{a^{-3} b^2}{a^4 b}\right)^{-4}$

7. $\left(\dfrac{x^{-2} y^3}{x^2 y^4}\right)^{-3}$

8. $\left(\dfrac{b^{-3} c^5}{b^4 c^{-2}}\right)^{-2}$

9. $\left(\dfrac{a^3}{x^5}\right)^{-5}$

10. $\left(\dfrac{a^5 c^2}{c^4 a^{-3}}\right)^{-1}$

Review of Negative Exponents

1. x^{-4}

2. $\dfrac{a^3}{a^{-2}}$

3. $\dfrac{a^4 b^9}{a^{-2} b^{-5}}$

4. $\dfrac{b^{-6}}{c^{-5}}$

5. $\dfrac{4^{-2} a^3}{2 a^{-2}}$

6. $\dfrac{a^3 b^2 c}{a^{-3} b^2 c^{-3}}$

7. $\left(\dfrac{a^5}{a^2}\right)^3$

8. $\left(\dfrac{2x}{y^4}\right)^{-1}$

9. $\left(\dfrac{a^2 x^{-3}}{a^{-2} x^4}\right)^{-2}$

10. $\dfrac{5^{-2}}{x^{-2}}$

11. $\left(\dfrac{x^2 y^{-3}}{y^3 x}\right)^{-3}$

12. $\left(\dfrac{a^2}{a^3}\right)^4$

13. $\dfrac{a^{-2} b^2}{4^2 b^{-2}}$

14. 2^{-4}

15. $\dfrac{a^{-2} y^{-2} b^3}{a^2 y^3 b^{-2}}$

16. $\left(\dfrac{a^{-4}}{a^3}\right)^2$

Chapter 4 Review

1. $x+5 \overline{) x^2 + 12x + 35}$

2. $\dfrac{5a^5b^4}{-a^4b}$

3. $\dfrac{2^{-3}y}{4y^{-2}}$

4. $-2(a^2b)^0$

5. $\dfrac{24x^2y + 32x^3y^2}{-4x^2y}$

6. $x-5 \overline{) 3x^2 - 14x - 5}$

7. $\left(\dfrac{a^3}{a^5}\right)^4$

8. $\dfrac{20x^5}{-4x}$

9. 3^{-3}

10. $\dfrac{-x^3}{x^7}$

11. $\dfrac{a^5b^2}{a^{-3}b^{-5}}$

12. $\dfrac{3x^3+16x^2-64}{x+4}$

13. $x+2 \overline{) 3x^3 + 11x^2 + 8x - 4}$

14. $\dfrac{xy^{-4}}{x^{-3}y^2z^2}$

15. $\left(\dfrac{x^{-2}y^6}{x^3y^{-2}}\right)^{-1}$

58

Cumulative Review 3

Multiply:

1. $5x(3x - 10)$

2. $(x + 9)(x + 4)$

3. $(2x^2y^3)^3$

4. $\quad 3x^2 - 7x + 1$
   $\quad \underline{4x - 2}$

5. $(6x^2)(-3x^3y^2)$

Add or Subtract:

6. $20a^2 - 14a - 8 - 9a + 16 + 2a^2 + 9 - 4a$

7. $(9x^2 + 9x + 10) - (8x^2 + 3x - 11)$

8. Subtract:
   $\quad x^3 + 5x^2 + 3x - 18$
   $\quad \underline{4x^3 - 7x^2 - 9x + 13}$

9. Subtract: $\quad 4x^2 - 7x + 2$
   $\quad \quad \quad \quad \underline{6x^2 - 4x - 8}$

Name the underlined:

10. $\underline{2x} - y$ _____

11. Identify the number: 0 _____

Name the property:

12. $x = 10$
    $6x - 23 = 37$
    $6(10) - 23 = 37$ _____

13. $2x + 0 = 2x$ _____

Simplify:

14. $4(7 - 13)$

15. $8 + (4 - 3\{2 + [9 - 3] - 10\} + 8) - 13$

16. $6^2 - 4^2 + 2 \times 3^2 - 12$

17. Is "8" the solutuion to the equation
    $4x + 3 - 2x = 16 + 5x - 36$?

Chapter 5
Solving Equations

Removing Parentheses - before solving an equation, all parentheses in the equation must first be removed. As a reminder, a (+) sign in front of a ( ) changes <u>nothing</u>, a (-) sign in front of a ( ) changes <u>all</u> the signs <u>inside</u> the ( ), and a term next to a ( ) requires <u>everything</u> inside the parentheses to be <u>multiplied</u> by that term.

Examples

a. $3(x + 4) = 16 - (2x - 3)$ b. $5x + (8x + 3) = 6 - 5(2x - 3)$
$3x + 12 = 16 - 2x + 3$ $5x + 8x + 3 = 6 - 10x + 15$

Remove the parentheses:

1. $5(x + 9) = 7 - (x - 4)$
2. $3 + (x - 9) = 2(x + 6)$
3. $5 - (x + 5) = 3(4x - 1)$
4. $4(2x + 6) = 2 + (x - 6)$
5. $5 + (6x + 4) = 4(2x - 9)$
6. $7(4x - 6) = 3 + (2x + 1)$
7. $2x + 3(x - 1) = 6(x + 4)$
8. $4 - (x - 5) = 2 - (x + 6)$
9. $5x + 2(x - 1) = 14 + 3(x+2)$

10. $6x - (x + 9) = 9 + (3x - 1)$
11. $6x + (4x - 1) = 15 - (3x + 6)$
12. $8x - 2(x + 6) = 11 + 3(2x - 3)$
13. $4 + 7(2x - 6) = 7 - 4(x + 3)$
14. $9 + (3x - 4) = 15 - (2x - 2)$
15. $3 - (x + 1) = 2(x + 6)$
16. $5(x - 4) = 3 + (x - 7)$
17. $4 - (2x - 3) = 9 - (x + 8)$
18. $2 + (x + 6) = 3 + (4x - 1)$

19. $-3(x - 1) = 15$
20. $-4(2x + 9) = 20$
21. $-4(x + 5) = 2 + (x + 2)$
22. $-5 - (3x - 1) = 4(2x + 5)$
23. $-10 + (x + 6) = 4 - (2x - 3)$
24. $4x + 6(2x - 7) = 14 + (x - 1)$
25. $10 - 4(x + 3) = x + 2(x - 2)$
26. $3(x + 5) = 2(x + 4)$
27. $-7 - (4x - 5) = -2 + (x + 1)$

Combining Like Terms - before solving an equation, the like terms on **each side** of the equation must be combined.

Example
$$11x - 19 - 4x = 19 + 6x - 31$$
$$7x - 19 = -12 + 6x$$

Combing the like terms:

1. $4x - 7 + 3x = 14 + 2x - 11$

2. $5x + 8 - 2x = 19 + 4x + 5$

3. $7x - 10 + 3x = 21 - 4x + 5$

4. $26 + 4x - 3 = 6x + 7 - 2x$

5. $27 - 8x + 11 = 10x - 11 + 3x$

6. $16 + 4x + 9 = 15x + 11 - 3x$

7. $12x + x - 3 = 5x - 14 + 10$

8. $14 + 7 + 5x = 6x + x - 9$

9. $6 - 3x + 2x = 5x + 21 - 3$

10. $7x - 2x = 21 + 3$

11. $8x + 4x = 14 - 2$

12. $15 - 3 = 8x - 4x$

13. $27 + 10 = 5x + x$

14. $6x - 3 + 2x + 10 = 15 - 2x + 4$

15. $8x + 9 - 3x - 15 = 11x + 21 - 4x$

16. $4x - 3 + 2x + 9 = 16 - 2x + 9$

17. $15 - x + 2 - 2x = 0$

18. $14 - 2x - 3 = 6x + 1 - 3x + 2x$

19. $6x + 3x - x = 26 - 4 + 7$

20. $19 + 2 - 5 = 8x + 3x - 2x$

21. $6 - 5 - 4 = 3x - 2x + x$

22. $8x + 2x + x = 14 - 3 - 20$

23. $6x + x - 10 = 4x - x + 3$

24. $14 - x + 4 = 12 + 2x - 3$

One step Equations - Addition and Subtraction - when the parentheses have been removed and the like terms on each side of the equation have been combined, this general rule can be used in all simple equations: <u>Move the known values (the numbers) to one side of the equation and move the unknown values (the variables) to the other side by using the opposite operation.</u> (Addition and subtraction are opposites; multiplication and division are opposites.)

Examples

$$
\begin{array}{ll}
\text{a.} \quad x + 23 = 10 & \text{b.} \quad x - 15 = -12 \\
 -23 = -23 & +15 = +15 \\
x = -13 & x = +3
\end{array}
$$

Solve:

1. $x + 11 = 19$
2. $x + 18 = 16$
3. $x + 15 = 29$
4. $x + 3 = -14$
5. $x + 9 = -31$
6. $x + 25 = -40$
7. $x + 29 = 37$
8. $x + 47 = -20$
9. $x + 24 = 9$

10. $x - 97 = 3$
11. $x + 20 = 18$
12. $x + 19 = -35$
13. $x + 6 = 20$
14. $x - 9 = 16$
15. $x - 12 = 7$
16. $x - 13 = -5$
17. $x - 20 = -11$
18. $x - 15 = 21$

19. $x - 3 = -11$
20. $x - 16 = 5$
21. $x - 12 = 18$
22. $x - 9 = -5$
23. $x - 31 = 16$
24. $x - 7 = 29$
25. $x - 22 = -30$
26. $x - 2 = -7$
27. $x + 13 = -18$

28. $x - 5 = 14$
29. $x + 23 = 41$
30. $x - 22 = 11$
31. $x + 32 = 45$
32. $x - 17 = -30$
33. $x - 14 = -6$
34. $x + 20 = -29$
35. $x - 1 = 21$
36. $x + 50 = 32$

One Step Equations - Multiplication - if the equation contains multiplication, (5a = 30), move the known value (5) to the other side of the equation be dividing both sides by the number being moved. (5)

Examples:

a. $6x = 30$
$\frac{6x}{6} = \frac{30}{6}$
$x = 5$

b. $8x = 20$
$\frac{8x}{8} = \frac{20}{8}$
$x = 2\frac{1}{2}$

c. $-4x = 28$
$\frac{-4x}{-4} = \frac{28}{-4}$
$x = -7$

d. $-7x = 3$
$\frac{-7x}{-7} = \frac{3}{-7}$
$x = -\frac{3}{7}$

Solve:

1. $5x = 50$
2. $2x = 16$
3. $3x = 27$
4. $4x = 24$
5. $7x = 35$
6. $8x = 40$
7. $6x = 18$
8. $4x = -20$
9. $9x = -36$
10. $5x = -45$
11. $3x = 99$
12. $10x = -70$
13. $12x = -48$
14. $6x = -36$
15. $3x = -33$
16. $2x = -10$
17. $8x = 12$
18. $4x = 18$
19. $10x = 25$
20. $6x = 39$
21. $11x = 16$
22. $9x = 30$
23. $5x = 19$
24. $2x = 13$
25. $7x = 31$
26. $3x = 25$
27. $-7x = 14$
28. $-4x = 32$
29. $-5x = -15$
30. $-10x = 35$
31. $-8x = -48$
32. $-6x = 60$
33. $-4x = -24$
34. $-12x = 26$
35. $-5x = 70$
36. $-7x = -35$
37. $-11x = 33$
38. $6x = 5$
39. $5x = 2$
40. $8x = 3$

One Step Equations - Division - if there is division in the problem, ($\frac{x}{5}$ is division), move the known value (5) to the other side of the equation by multiplying both sides of the equation by that known value (5). If the division problem contains two integers (a fraction, $\frac{3x}{5}$), multiply both sides by the multiplicative inverse.

Examples

a. $\frac{x}{3} = 7$
$\frac{3}{1} \cdot \frac{x}{3} = 7 \cdot \frac{3}{1}$
$\frac{x}{1} = \frac{21}{1}$
$x = 21$

b. $\frac{x}{7} = -4$
$\frac{7}{1} \cdot \frac{x}{7} = -4 \cdot \frac{7}{1}$
$\frac{x}{1} = \frac{-28}{1}$
$x = -28$

c. $\frac{3x}{4} = 15$
$\frac{4}{3} \cdot \frac{3x}{4} = \frac{15}{1} \cdot \frac{4}{3}$
$\frac{x}{1} = \frac{60}{3}$
$x = 20$

d. $\frac{2x}{3} = -6$
$\frac{3}{2} \cdot \frac{2x}{3} = \frac{-6}{1} \cdot \frac{3}{2}$
$\frac{x}{1} = \frac{-18}{1}$
$x = -18$

Solve:

1. $\frac{x}{2} = 7$
2. $\frac{x}{5} = 3$
3. $\frac{x}{6} = 9$
4. $\frac{x}{3} = 4$
5. $\frac{x}{8} = -3$
6. $\frac{x}{6} = -9$

7. $\frac{x}{5} = -7$
8. $\frac{x}{11} = -4$
9. $\frac{x}{9} = -2$
10. $\frac{2x}{5} = 6$
11. $\frac{3x}{4} = 9$
12. $\frac{5x}{6} = 15$

13. $\frac{5x}{2} = 20$
14. $\frac{4x}{3} = 12$
15. $\frac{4x}{7} = 20$
16. $\frac{2x}{7} = -12$
17. $\frac{5x}{4} = -25$
18. $\frac{6x}{5} = -18$

19. $\frac{x}{5} = -4$
20. $\frac{x}{10} = 6$
21. $\frac{3x}{4} = 33$
22. $\frac{2x}{7} = -18$
23. $\frac{x}{10} = -5$
24. $\frac{4x}{5} = -16$

25. $\frac{x}{9} = -3$
26. $\frac{x}{7} = 4$
27. $\frac{6x}{7} = 24$
28. $\frac{5x}{2} = -30$
29. $\frac{x}{5} = -6$
30. $\frac{7x}{10} = -28$

One Step Equations - Negative X - when (-x) appears in the solution, the negative x must be changed to positive x. When the sign of (-x) is changed, <u>every</u> sign in the equation must also be changed. Normally, this can be the last step in solving the equation.

Examples

a. $15 - x = +23$
$-15 -15$
$-x = +8$
$+x = -8$

b. $-x - 7 = -18$
$+7 +7$
$-x = -11$
$+x = +11$

Solve:

1. $10 - x = 19$
2. $3 - x = 12$
3. $8 - x = 5$
4. $15 - x = 6$
5. $16 - x = -25$
6. $-7 - x = 2$
7. $12 - x = -9$
8. $-8 - x = 12$

9. $9 - x = 15$
10. $-x + 10 = 3$
11. $-x - 16 = 10$
12. $-x + 7 = -6$
13. $-x - 15 = -3$
14. $-x + 5 = 2$
15. $14 - x = 13$
16. $12 - x = -5$

17. $-x - 6 = 12$
18. $-3 - x = 16$
19. $-17 - x = -11$
20. $-x + 13 = 2$
21. $-x - 10 = -6$
22. $16 - x = -16$
23. $7 - x = 10$
24. $-x + 7 = -12$

Two Step Equations - these problems have two numbers that must be moved to the other side of the equation, making two separate steps necessary. Start with addition/subtraction before multiplication/division.

Examples

a. $\begin{aligned} 3x - 5 &= 16 \\ +5 &+ 5 \\ \hline 3x &= 21 \\ \overline{3} & \overline{3} \\ x &= 7 \end{aligned}$

b. $\begin{aligned} \tfrac{x}{4} - 6 &= 3 \\ +6 &= +6 \\ \hline \tfrac{x}{4} &= +9 \\ \tfrac{4}{1} \cdot \tfrac{x}{4} &= \tfrac{4}{1} \cdot \tfrac{9}{1} \\ x &= 36 \end{aligned}$

c. $\begin{aligned} \tfrac{2x}{3} + 7 &= 13 \\ -7 &- 7 \\ \hline \tfrac{2x}{3} &= 6 \\ \tfrac{3}{2} \cdot \tfrac{2x}{3} &= \tfrac{3}{2} \cdot \tfrac{6}{1} \\ x &= 9 \end{aligned}$

Solve:

1. $2x + 3 = 11$
2. $2x - 8 = 6$
3. $3x + 8 = 17$
4. $4x - 10 = 18$
5. $5x + 15 = 40$
6. $6x - 5 = 55$
7. $4x + 32 = 48$
8. $8x + 27 = 67$
9. $\tfrac{x}{5} + 3 = 8$
10. $\tfrac{x}{3} + 4 = -2$
11. $\tfrac{x}{8} - 7 = 3$
12. $\tfrac{x}{10} - 6 = -10$
13. $\tfrac{x}{6} + 9 = 6$
14. $\tfrac{x}{4} - 5 = 1$
15. $\tfrac{x}{7} - 6 = -8$
16. $\tfrac{x}{2} + 10 = 14$
17. $\tfrac{2x}{5} + 9 = -1$
18. $\tfrac{4x}{3} - 2 = 10$
19. $\tfrac{2x}{7} - 8 = -4$
20. $\tfrac{5x}{3} + 4 = 24$
21. $\tfrac{4x}{7} - 6 = 10$
22. $\tfrac{6x}{5} + 5 = 23$
23. $\tfrac{4x}{7} + 7 = +3$
24. $\tfrac{5x}{6} - 3 = -28$
25. $4x + 16 = 36$
26. $\tfrac{x}{4} - 6 = -10$
27. $10x - 4 = -24$
28. $8x - 7 = 33$

Equations With One Fraction - a term that looks like division ($\frac{x}{6}$) can also be written as a fraction ($\frac{1}{6}x$). To change a fraction to "1", multiply the fraction by its multiplicative inverse. Multiply the other side of the equation by the same multiplicative inverse.

Examples

a. $\frac{1}{6}x = 5$
$\frac{6}{1} \cdot \frac{1}{6}x = \frac{5}{1} \cdot \frac{6}{1}$
$x = 30$

b. $\frac{3}{4}x = 12$
$\frac{4}{3} \cdot \frac{3}{4}x = \frac{12}{1} \cdot \frac{4}{3}$
$x = 16$

c. $\frac{2}{5}x - 3 = 11$
$\phantom{\frac{2}{5}x}+3 +3$
$\frac{2}{5}x = 14$
$\frac{5}{2} \cdot \frac{2}{5}x = 14 \cdot \frac{5}{2}$
$x = 35$

Solve:

1. $\frac{1}{3}x = 4$
2. $\frac{1}{2}x = 7$
3. $\frac{2}{5}x = 12$
4. $\frac{3}{2}x = 9$
5. $\frac{1}{6}x + 2 = 8$

6. $\frac{7}{5}x - 3 = 11$
7. $\frac{6}{5}x = -24$
8. $\frac{1}{4}x - 5 = 3$
9. $\frac{1}{5}x = -2$
10. $\frac{4}{5}x = 20$

11. $\frac{1}{5}x - 8 = -4$
12. $\frac{2}{3}x = 6$
13. $\frac{5}{4}x = 30$
14. $\frac{7}{8}x + 2 = 16$
15. $\frac{7}{5}x - 9 = -23$

16. $\frac{1}{8}x + 4 = 7$
17. $\frac{3}{5}x = 15$
18. $\frac{3}{8}x - 6 = -15$
19. $\frac{4}{7}x + 10 = 14$
20. $\frac{1}{5}x = -3$

Two Step Equations With Four Terms - move the smaller number of x's to the side of the larger number of x's. Move the integers to the other side. Note that in one of the examples the solution is a fraction.

Examples

a. $\begin{aligned} 5x - 3 &= 7x + 12 \\ \underline{-5x - 12} & \underline{-5x - 12} \\ \frac{-15}{2} & \frac{2x}{2} \\ -7\frac{1}{2} &= x \end{aligned}$

b. $\begin{aligned} 6x - 3 &= 4x + 9 \\ \underline{-4x + 3} & \underline{-4x + 3} \\ \frac{2x}{2} &= \frac{+12}{2} \\ x &= 6 \end{aligned}$

Solve:

1. $4x + 9 = 6x + 7$

2. $8x + 18 = 10x - 10$

3. $5x - 3 = 8x + 15$

4. $6x - 11 = 10x + 17$

5. $4x + 13 = 9x - 27$

6. $11x + 2 = 14x - 10$

7. $7x - 3 = 13x + 33$

8. $6x + 12 = 4x + 22$

9. $8x - 3 = 4x + 21$

10. $8x - 1 = 2x - 37$

11. $10x + 3 = 7x - 39$

12. $7x - 13 = 2x - 18$

13. $12x + 20 = 2x + 40$

14. $5x - 18 = x - 2$

15. $8x + 2 = 7x - 1$

16. $4x - 3 = 3x + 5$

17. $5x + 9 = 4x + 16$

18. $10x - 3 = 9x - 7$

19. $8x + 2 = 9x + 6$

20. $7x - 3 = 8x + 9$

21. $5x + 15 = 6x - 9$

22. $7x - 11 = 6x + 3$

23. $6x - 3 = 4x + 10$

24. $8x + 9 = 3x - 25$

Multi-Term Equations - if there is more than one like term on any side of the equation, those like terms must be combined before you start to solve the equation.

Example

$$8x - 14 + 23 = 2x - 16 + 3x$$
$$8x + 9 = 5x - 16$$
$$\underline{-5x - 9 \quad -5x - 9}$$
$$\frac{3x}{3} = \frac{-25}{3}$$
$$x = -\frac{25}{3} \text{ or } -8\frac{1}{3}$$

(the -14 & the 23 have been combined of the left side of the equation, and the 2x & the 3x have been combined on the right side of the equation.)

Solve:

1. $10x - 3 + 5 = 8x + 26 + 4x$

2. $4x - 7 + 3x = 14 + 2x - 11$

3. $19 + 3x - 6 = 2x - 31 + 3x$

4. $5x + 8 - 2x = 19 + 4x + 5$

5. $15 + 6x - 10 = x - 18 + 3x$

6. $26 + 4x - 3 = 7x + 7 - 2x$

7. $7x - 16 + 2x = 25 + 5x - 3$

8. $12x + x - 3 = 5x - 14 + 10$

9. $7x - 14 - 6 = 6x + 5x + 18$

10. $14 + 7 + 5x = 6x + x - 9$

11. $8x - 3x + 2x = 27 + 11 - 9$

12. $6 - 3x + 2x = 5x + 21 - 3$

13. $6x - 14 - 3x = 14 - 2x + 8$

14. $7x - 2x = 21 + 3$

15. $35 - 9x + 14 = 30 - 9x + 3x$

Equations With Parentheses - all parentheses must be removed before solving the equation. After removing the parentheses, combine any like terms that are on the <u>same</u> sides of the equation.

Example

$$4 + 3(x - 2) = 6x - (2x + 9) \quad \{\text{remove the parentheses}\}$$
$$4 + 3x - 6 = 6x - 2x - 9 \quad \{\text{combine the like terms}\}$$
$$3x - 2 = 4x - 9 \quad \{\text{solve the equation}\}$$
$$\underline{-3x + 9 \quad -3x + 9}$$
$$+ 7 = x$$

Solve:

1. $10 + 5(x - 4) = 8x - (5x + 4)$

2. $20 + 3(x - 8) = 19 - (3x - 17)$

3. $5(x + 9) = 7 - (x - 4)$

4. $3(x - 9) = 2(x + 6)$

5. $10x - 3(x - 8) = 19 - (3x - 17)$

6. $4(x + 2) + 10 = 6x - (x + 9)$

7. $5(2x + 6) = 2(x - 6)$

8. $3x + (13 - x) = 15 - (9 - x)$

9. $5(6x + 4) = 4(2x - 9)$

10. $5 + 3(2x - 3) = 6x - (x - 3)$

11. $6x - 3(x + 9) = 9 + 2(3x - 1)$

12. $8x + 3(x + 9) = 27 - 3(x + 6)$

13. $6x + (4x - 1) = 15 - (3x + 6)$

14. $25 - 9(x + 1) = 17 + 2(4x + 7)$

15. $3(x + 1) = 11 + 3(2x - 3)$

Absolute Value Equations - absolute value tells us that $|+7| = 7$ and $|-7| = 7$. In the equation: $|x| = 7$, the x can be replaced with a (+ 7) or a (- 7). Therefore, $x = +7$ or $x = -7$

Examples

a. $|x| = 8$
   $x = +8$; $x = -8$

b. $|3x| = 27$
   $\dfrac{3x}{3} = \dfrac{+27}{3}$; $\dfrac{3x}{3} = \dfrac{-27}{3}$
   $x = +9$; $x = -9$

Solve:

1. $|x| = 3$
2. $|x| = 9$
3. $|x| = 15$
4. $|x| = 4$
5. $|x| = 10$
6. $|x| = 12$
7. $|x| = 5$
8. $|x| = 6$

9. $|4x| = 16$
10. $|9x| = 27$
11. $|8x| = 40$
12. $|10x| = 20$
13. $|7x| = 28$
14. $|3x| = 30$
15. $|5x| = 45$
16. $|9x| = 18$

17. $|4x| = 17$
18. $|11x| = 7$
19. $|7x| = 12$
20. $|9x| = 11$
21. $|3x| = 19$
22. $|12x| = 21$
23. $|3x| = 14$
24. $|5x| = 2$

25. $|x| = 8$
26. $|6x| = 8$
27. $|13x| = 3$
28. $|x| = 20$
29. $|14x| = 20$
30. $|7x| = 63$
31. $|3x| = 9$
32. $|10x| = 50$

More Absolute Value Equations - a polynomial inside the <u>absolute value bars</u> can be replaced with (and is equal to) the positive and negative number on the other side of the equation.

Examples

a. $|x + 3| = 6$
$x + 3 = 6$; $x + 3 = -6$
$ -3 -3$; $ -3 = -3$
$x = 3$; $x = -9$

b. $|3x - 5| = 4$
$3x - 5 = 4$; $3x - 5 = -4$
$ +5 +5$; $ +5 +5$
$\frac{3x}{3} = \frac{9}{3}$; $\frac{3x}{3} = \frac{1}{3}$
$x = 3$; $x = \frac{1}{3}$

Solve:

1. $|x + 2| = 6$

2. $|x + 3| = 10$

3. $|x - 1| = 9$

4. $|x - 4| = 3$

5. $|x - 7| = 7$

6. $|x + 6| = 4$

7. $|x + 5| = 8$

8. $|2x + 1| = 2$

9. $|3x - 2| = 1$

10. $|2x - 5| = 5$

11. $|3x + 4| = 0$

12. $|4x - 1| = 3$

13. $|5x + 4| = 6$

14. $|4x - 3| = 4$

15. $|x + 5| = 10$

Chapter 5 Review
Solve:
1. $7x = -21$

2. $4x - 3 = 3x + 7$

3. $|x| = 9$

4. $12 - x = 7$

5. $\dfrac{x}{6} = 5$

6. $7x + 2 - 3x = 21 - 2x + 13$

7. $x + 19 = -6$

8. $|4x| = 24$

9. $\dfrac{x}{6} + 2 = 9$

10. $|4x - 3| = 17$

11. $\dfrac{5x}{2} = 15$

12. $x - 9 = -1$

13. $4x + 10 = 34$

14. $-8x = 13$

15. $8x - 5 = 3x + 27$

16. $\dfrac{4x}{3} - 6 = 6$

17. Remove the parentheses:
$4 - 3(2x + 1) = 6x + (4x - 3)$

18. Combine the like terms:
$7x - 3 + 2x + 1 = 4x - 19 + x$

Cumulative Review 4

1. $\dfrac{3x^4y^3}{-xy^2}$

2. $\dfrac{18x^2y^3 - 6x^3y^2}{6x^2y^2}$

3. $x-3 \,\overline{\big)\, 2x^2 \quad -x \quad -15}$

4. $\left(\dfrac{x^6}{x^9}\right)^3$

5. $\dfrac{a^{-4}}{a^3}$

Multiply:

6. $(x^5y^4)^3$

7. $\begin{array}{r} x - 5 \\ \underline{x + 8} \end{array}$

8. $\begin{array}{r} 7x^2 + 4x - 15 \\ \underline{-3x} \end{array}$

9. $(4x - 7)(3x - 2)$

Add or Subtract:

10. $(5x^2 - 9x) - (7x + 19)$

11. $\begin{array}{r} 6x^2 - 2x - 9 \\ 5x^2 + 7x - 2 \\ \underline{-9x^2 - 3x + 5} \end{array}$

Name the underlined:

12. 6x + <u>17</u> _____

13. <u>| - 6 |</u> _____

Simplify:

14. $5 - (5 + \{4 - 3[8 + 2] - 9\} + 2) - 14$

15. Find the value of Z
    $Z = 3x^2 - 25; \quad x = 4$

Chapter 6
Inequalities

An inequality is an equation with an <u>inequality sign</u> instead of an equal sign.

To <u>read</u> an inequality the following must be known
1. > means "is greater than" and < means "is less than" when read from left to right.
2. ≥ means "is greater than or equal to" and ≤ means "is less than or equal to" when read from left to right.
3. The term of the **point** side of the sign is **less** than the term on the **open** side of the sign. (The open side is greater than the point side.)
4. When reading inequalities, always begin with the "x".

Examples
a. $x > 6$ is written or read as: **X is greater than 6.**
b. $-2 \leq x$ is written or read as: **X is greater than or equal to -2.**
c. $x < 4$ is written or read as: **X is less than 4.**

Write the meaning of the inequality

1. $x > 5$ _____

2. $x \leq -2$ _____

3. $7 \leq x$ _____

4. $x < -1$ _____

5. $x \leq -5$ _____

6. $-11 \leq x$ _____

7. $4 > x$ _____

8. $3 < x$ _____

9. $x > -4$ _____

10. $x \leq 4$ _____

11. $6 > x$ _____

Graphing Inequalities, the Number Line - the solution to an inequality can be put on a number line. This is called "graphing the solution". The following must be known:

1. The number line is a horizontal line with a zero marked on it. All positive integers are to the right of the zero, all negative integers are to the left of the zero.
2. When graphing the solution, put the "solution" on the number line and draw an arrow from than number, <u>greater than</u> to the **right** and <u>less than</u> to the **left**.
3. When the solution is <u>greater than or less than</u>, the arrow does **not touch** the number, when the solution is <u>greater than or equal to</u> or <u>less than or equal to</u>, the arrow **does touch** the line.

Examples

a. x > -2, (X is greater than -2.)

b. 4 ≥ x. (X is less than or equal to 4.)

Put the solution on the number line

1. x > 4

2. x ≤ 7

3. x ≥ -4

4. 5 < x

5. x ≤ -3

6. -1 < x

7. 0 < x

8. x ≤ 0

9. 7 > x

10. x ≥ 6

11. x < -3

12. -7 ≤ x

13. 5 > x

14. x > -2

15. x ≤ 5

16. 2 ≥ x

17. -3 > x

18. x ≤ 2

Inequalities, The Solution set - the solution to inequalities can be put in set form: x = { }. The solution set is the written solution to the inequality.

Example

$x \leq -3$

x = {all real numbers less than or equal to 3}

Graph the solution and give the solution in set form

1. $x < 2$

   x = { }

2. $5 < x$

   x = { }

3. $x \leq 6$

   x = { }

4. $4 < x$

   x = { }

5. $x \geq -2$

   x = { }

6. $x > 6$

   x = { }

7. $-5 < x$

   x = { }

Inequalities, the Solution Set Continued

8. -1 > x ———|———
              0

   x = {   }

9. 5 ≤ x ———|———
             0

   x = {   }

10. -2 > x ———|———
               0

    x = {   }

11. x ≤ 5 ———|———
              0

    x = {   }

12. 7 > x ———|———
              0

    x = {   }

13. x ≤ 0 ———|———
              0

    x = {   }

14. 2 ≥ x ———|———
              0

    x = {   }

15. -1 ≤ x ———|———
               0

    x = {   }

16. x < 1 ———|———
              0

    x = {   }

Inequalities, Negative x as the Solution - it is possible than the solution will be a negative x. When this occurs, the negative x and all other signs must be changed, as well as the direction of the inequality sign.

Examples

a. $-x > 4$

   $x < -4$

   The (-x), (4), and (>) have all been changed.

b. $-2x \leq -6$

   $\dfrac{-2x}{-2} \leq \dfrac{-6}{-2}$

   $x \geq 3$

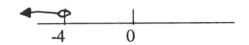

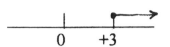

Solve and put the solution on the number line.

1. $-x > -6$

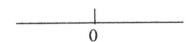

2. $-x \leq -3$

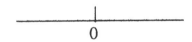

3. $-3x \geq -12$

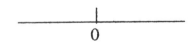

4. $-5x < 15$

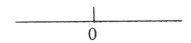

5. $6 - x > 7$

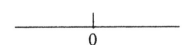

6. $3 - x \leq -2$

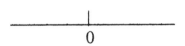

7. $-x < -6$

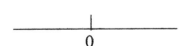

8. $7 - x \geq 2$

Solving Inequalities, All Types - any equation can also be an inequality. The same steps taken to solve equations are also taken when solving inequalities.

Example

$$4(x - 2) \geq 2x - 16$$
$$4x - 8 \geq 2x - 16$$
$$\underline{-2x + 8 -2x + 8}$$
$$\frac{2x}{2} \geq \frac{-8}{2}$$
$$x \geq -4$$

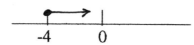

x = {all the real numbers greater than or equal to -4}

Solve, put on the number line, and give the solution in set form

1. $x + 1 < 4$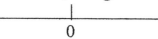

   x = {   }

2. $x - 2 > -3$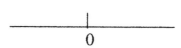

   x = {   }

3. $5x \geq 30$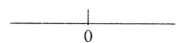

   x = {   }

4. $3x - 7 \leq 5$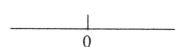

   x = {   }

Solving Inequalities, All Types Continued

5. $9x + 4 > 6x + 25$ ──────|──────
                                       0

    x = {   }

6. $6 - x \leq 4$ ──────|──────
                                      0

    x = {   }

7. $\dfrac{2x}{3} < -4$ ──────|──────
                                       0

    x = {   }

8. $5(2x - 3) > 3(3x + 4)$ ──────|──────
                                                  0

    x = {   }

9. $x + 23 \leq 19$ ──────|──────
                                      0

    x = {   }

Compound Inequalities - have two inequality symbols and each inequality can be read separately, beginning with the "x".

Example
$-2 < x + 1 \geq -3$ is read as:
"x + 1" is greater than negative two and "x + 1" is greater than or equal to negative three."

Each inequality can be put on one number line, and the solution is all real numbers touched or crossed by both arrows.

Examples

a. $\quad 0 < x > 4$

{this example is already sloved}

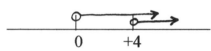

$\quad\quad 0 \quad\quad +4$

x = {all real numbers greater than four}

b. $\quad\quad -2 \leq x + 2 \leq 3$

{this example must first be solved for x}
$\quad\quad \underline{-2 \quad\quad -2 \quad -2}$
$\quad\quad -4 \leq \quad x \quad \leq 1$

$\quad\quad -4 \quad\quad 0 \; +1$

x = {all real numbers greater than or equal to -4 but less than or equal to +1}

Solve, put on the number line, and give the solution set

1. $-2 > x > 4$

    x = { $\quad\quad\quad\quad\quad\quad\quad\quad\quad\quad$ }

2. $-6 \leq x < 2$

    x = { $\quad\quad\quad\quad\quad\quad\quad\quad\quad\quad$ }

3. $-1 \leq x \geq 3$

    x = { $\quad\quad\quad\quad\quad\quad\quad\quad\quad\quad$ }

4. $-3 > x < 4$

    x = { $\quad\quad\quad\quad\quad\quad\quad\quad\quad\quad$ }

5. $-5 > x + 4 > -1$

    x = { $\quad\quad\quad\quad\quad\quad\quad\quad\quad\quad$ }

Compound Inequalities Continued

6. $-2 \geq 2x \geq 8$

   x = {   }

7. $-7 < x - 5 \leq 0$

   x = {   }

8. $15 > 3x < 3$

   x = {   }

9. $4 \leq x + 3 < -2$

   x = {   }

10. $-5 < 5x > 20$

    x = {   }

11. $-1 \geq x - 4 \geq -7$

    x = {   }

Review of Inequalities

Solve, put on the number line, and give the solution set

1. $5 > x - 3$

   x = {   }

2. $6x < 30$

   x = {   }

3. $-1 < x + 3 < 5$

   x = {   }

4. $\frac{x}{3} - 2 > 3$

   x = {   }

5. $4x - 3 \leq 3x + 1$

   x = {   }

6. $10 - x > 4$

   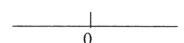

   x = {   }

Cumulative Review 5

Solve for x:

1. $5x = -35$

2. $7x - 23 = 5x + 11$

3. $|x| = 4$

4. $8 - x = -3$

5. $\dfrac{x}{8} = -2$

6. Remove the parentheses:
   $11 - 7(3x - 5) = 8x + (2x + 20)$

Simplify:

7. $\dfrac{6x^3 a^6}{4x^2 a^7}$

8. $x - 6 \overline{\smash{\big)}\, x^2 - 11x + 30}$

9. $-x(y^3 z^4)^0 =$

10. $\left(\dfrac{a^2 b^{-5}}{a^{-3} b^2}\right)^{-2}$

Multiply:

11. $2x^2 + 9x - 3$
    $\underline{5x - 1}$

12. $7xy^2(3xy + 4x^2 y^2 - 9x^3 y^3)$

Add or Subtract:

13. $(10x^2 - 4x + 2) + (3x^2 + 8x + 6)$

14. Subtract: $5ab + 3bc - 2ac$
    $\underline{3ab - 4bc - 8ac}$

Name the property:

15. $2 + (7 + 11) = (2 + 7) + 11$ _____

16. $\dfrac{7x}{7} = \dfrac{-63}{7}$ _____

Chapter 7 & 8
Factoring

Factors - are terms, that when multiplied together, give you a product. **Factoring** is going from a product to its' factors. For example, the factors of "15" are "3" and "5". When the factors 3 and 5 are multiplied, the product is 15. Therefore, multiplication and factoring are opposite procedures. When a polynomial has been factored, the factors can be checked by multiplication. The product (of the factors) should be the same as the original problem.

Examples
Check the factors by multiplication:

The polynomials:	$3x^3a - 12x^2a + 9xa$	$2x^2 + 3x - 9$
Are these the factors?	$3xa(x^2 - 4x + 3)$	$(x - 3)(2x + 3)$
Check by multiplication:	$3x^3a - 12x^2a + 9xa$	$2x^2 - 3x - 9$
Are the factors correct?	YES	NO (the middle term is "-3x".

(If the product of the factors is the same as the problem, YES, the factors are correct.)

Are the factors correct? Check them by multiplication.

1. $6x + 8$
   $2(3x + 4)$

2. $6c - 7cd$
   $c(6 - 7d)$

3. $3a^2 + 9a$
   $3a(a + 3)$

4. $6xy^2 - 9xy + 12xy^2$
   $3xy(y + 3 + 4x)$

5. $x^2 - 36$
   $(x - 6)(x - 6)$

6. $4x^2 + 11x + 6$
   $(4x + 2)(x + 3)$

7. $x^2 + 3x - 18$
   $(x - 3)(x + 6)$

8. $x^2 + 10x + 25$
   $(x + 5)^2$

9. $2x^2 - 9x + 10$
   $(2x - 5)(x - 2)$

10. $x^2 - x - 20$
    $(x - 4)(x + 5)$

11. $6x^2 + 15x - 2ax - 5a$
    $(3x - a)(2x - 5)$

12. $9x^2 - 49$
    $(3x - 7)(3x + 7)$

13. $2a^2 - ab - 8a + 4b$
    $(2a - b)(a - 4)$

14. $4x^2 + 4x - 15$
    $(2x + 5)(2x - 3)$

15. $x^4 - 1$
    $(x - 1)(x + 1)(x^2 + 1)$

16. $x^2 - x - 2$
    $(x + 2)(x - 1)$

Factoring, Common Factors - When factoring, the first thing to check for is a common factor. A common factor is a term (integer, variable, or both) that can be divided evenly into each term of the polynomial. To factor a problem that has a common factor:
1. determine the common factor that will divide evenly into each term
2. put that common factor under the polynomial
3. divide the common factor into each term in the polynomial
4. put the quotient inside one set of parentheses.

Examples

$3x + 12$ $6a + 5ab$ $4y^2 - 8y$ $12x^2y - 16xy^2 + 24y^3$
$3(x + 4)$ $a(6 + 5b)$ $4y(y - 2)$ $4y(3x^2 - 4xy + 6y^2)$

(The first term is the <u>common factor</u>, what is inside the parentheses is the other factor.)

Factor

1. $3x + 6$
2. $4x + 6$
3. $3x - 12$
4. $6x + 8$
5. $4x - 10$

6. $2x - 5xz$
7. $7y - 3yb$
8. $4a + 5ab$
9. $7b + 6bC$
10. $6c - 7cd$

11. $3a + 9a^2$
12. $5z + 20z^2$
13. $4y - 10y^2$
14. $6x + 8x^2$
15. $3y - 15y^2$

16. $12R - 16R^2 + 8R^3$
17. $a^2b - ab + ab^2$
18. $x^3y + 2x^2y^2 - xy^3$
19. $24a^2b - 3ab + 9ab^2$
20. $30rs - 20r^2s^2 + 40r^3s^3$

Review of Common Factors

1. $8x - 16$
2. $3xy + 4x$
3. $2x^2 + 8x$
4. $6a^3 + 3a^2 - 12a$

5. $3x + 12$
6. $2x + 5xz$
7. $7x^2 + 28x$
8. $6y^3 + 9y^2 + 15y$

9. $8x^2 - 2x$
10. $12x^3 + 8x^2 - 4x$
11. $8 + 8a$
12. $4bx - 5b$

13. $4y^2 - 14y$
14. $7y - 8ay$
15. $a^3b - 3a^2b^2 - ab^3$
16. $3a^3 - 75a$

Factoring, The Difference Between Two Perfect Squares - this is the second type of factoring problem. A "perfect square" is the product of a **term times itself**.

Examples of perfect squares are:

1. numbers: 1, 4, 9, 16, 25, 36, 49, 64, 81, 100, 121, 144, 169, 196...
2. variables: $a^2, b^4, c^6, d^8, x^{10}, y^{12}, z^{14}$.....
3. numbers & variables: $4x^2, 9y^4, 16z^6, 25a^8, 36b^{10}, 49c^{12}, 64a^{14}x^{16}, 81y^{18}b^{20}$,.....

The square root of a term is the number and/or variable that, when multiplied by itself, gives you that term. The square root of 4 is 2, the square root of 25 is 5, the square root of x^2 is x, the square root of $49y^6$ is $7y^3$.

Practice Problems
Fill in the blanks _____.
1. 5 is the square root of the perfect square _____.
2. _____ is the square root of the perfect square 100
3. y is the square root of the perfect square _____.
4. _____ is the square root of the perfect square a^2
5. 3a is the square root of the perfect square _____.
6. _____ is the square root of the perfect square $144x^2y^6$
7. $5b^5$ is the square root of the perfect square _____.
8. _____ is the square root of the perfect square $81y^6$
9. 4y is the square root of the perfect square _____.
10. _____ is the square root of the perfect square 36
11. 3y is the square root of the perfect square _____.
12. _____ is the square root of the perfect square $25s^2$
13. 7b is the square root of the perfect square _____.
14. _____ is the square root of the perfect square $49a^2$
15. $2c^2$ is the square root of the perfect square _____.
16. _____ is the square root of the perfect square $16y^2$
17. _____ is the square root of the perfect square $81y^2$
18. 6 is the square root of the perfect square _____.
19. 8x is the square root of the perfect square _____.
20. _____ is the square root of the perfect square a^4b^6
21. _____ is the square root of the perfect square $4b^{12}$
22. _____ is the square root of the perfect square $9a^2b^6c^{10}$
23. x^2y is the square root of the perfect square _____.
24. 10a is the square root of the perfect square _____.
25. _____ is the square root of the perfect square 121
26. abc is the square root of the perfect square _____.
27. 9b is the square root of the perfect square _____.

Factoring, The Difference Between Two Perfect Squares Continued - this type of problem always has two terms that are perfect squares, and they will be separated by a negative sign, thus the name "difference between two perfect squares".

The factors of this type of problem are:
1. the square root of the first term <u>added to</u> the square root of the second term.
2. the square root of the first term <u>subtracted by</u> the square root of the second term.

Examples

a. $4x^2 - 25$
$(2x-5)(2x+5)$

b. $16y^2 - 9$
$(4y-3)(4y+3)$

c. $z^2 - 36$
$(z-6)(z+6)$

d. $81x^2 - 4$
$(9x-2)(9x+2)$

Factor

1. $x^2 - 36$
2. $x^2 - 9$
3. $x^2 - 81$
4. $x^2 - 49$
5. $x^2 - 1$
6. $x^2 - 64$
7. $x^2 - 100$
8. $x^2 - 4$
9. $x^2 - 25$
10. $x^2 - 144$

11. $x^2 - 16$
12. $x^2 - 121$
13. $4x^2 - 1$
14. $4x^2 - 9$
15. $16x^2 - 1$
16. $25x^2 - 36$
17. $16x^2 - 9a^6$
18. $36x^2 - 49y^2$
19. $25x^4 - 4y^2$
20. $1 - 9x^2$

21. $1 - 25z^2$
22. $1 - 49y^8$
23. $1 - 4x^2$
24. $49x^2y^6 - 36a^8$
25. $a^2 - b^2$
26. $x^2 - y^2$
27. $a^2b^2 - x^2y^2$
28. $a^6 - b^6$
29. $x^4 - b^{10}$
30. $1 - R^6$

Factoring, Trinomials That Factor as a Binomial Square - this is the third type of factoring problem. A "binomial square" is a binomial that is squared: $(x-7)^2$. A problem that will fctor as a "binomial square" is: $4x^2 + 12x + 9$. Notice, the first and third terms are **positive** perfect squares.

To factor this type of problem
1. Put the square root of the first and third terms in a parentheses.
2. Separate them by the sign of the middle term of the trinomal to be factored.
3. Put the exponent "2" to the right of the parentheses.
4. You must check your answer by doubling the product of the two square roots, because not all problems that have positive perfect squares for their first and third terms will factor this way.

Examples

a. $x^2 - 12x + 36$
   $(x-6)^2$

b. $x^2 + 18x + 81$
   $(x+9)^2$

c. $4x^2 - 12x + 9$
   $(2x-3)^2$

Factor

1. $x^2 + 8x + 16$
2. $z^2 + 20z + 100$
3. $b^2 - 14b + 49$
4. $y^2 - 4y + 4$
5. $c^2 + 10c + 25$
6. $x^2 - 6x + 9$
7. $z^2 - 16z + 64$
8. $a^2 - 2a + 1$

9. $c^2 + 12c + 36$
10. $y^2 + 18y + 81$
11. $z^2 + 8z + 16$
12. $a^2 - 22a + 121$
13. $x^2 + 14x + 49$
14. $b^2 - 16b + 64$
15. $9x^2 + 12x + 4$
16. $4x^2 - 28x + 49$

17. $16a^2 + 40a + 25$
18. $25y^2 + 20y + 4$
19. $49a^2 - 84a + 36$
20. $64x^2 - 112x + 49$
21. $36a^2 + 12a + 1$
22. $1 - 10x + 25x^2$
23. $1 + 8x + 16x^2$
24. $1 + 14x + 49x^2$

Review of Factoring:
1. Common factors
2. Difference between two perfect squares
3. Trinomials that factor as a binomial square

1. $8x - 16$

2. $x^2 - 16$

3. $x^2 + 10x + 25$

4. $2m - 3mn$

5. $9x^2 - 4$

6. $x^2 - 14x + 49$

7. $x^2 - 81$

8. $x^2 + 18x + 81$

9. $6x + 9x^2$

10. $12ab - 4b$

11. $x^2 - 25$

12. $10a^2 - 15a + 20$

13. $9x^2 - 49$

14. $6x^3 - 12x^2 + 3x$

15. $4x^2 - 1$

16. $x^2 - y^2$

17. $x^2 + 2x + 1$

18. $18a^2b - 15ab + 12ab^2$

19. $25x^2 - 9$

20. $3x^2 - 6x + 9$

21. $10xy^2 - 20xy - 30x^2y$

22. $64x^4 - 49y^2$

23. $x^2 - 8xy + 16y^2$

24. $15x^3 - 20x^2 + 10x$

25. $x^2 - 25y^2z^8$

26. $a^2 + 2ab + b^2$

27. $25x^2 - 30x + 9$

28. $100a^2 - 81b^2c^6$

29. $9x^2 + 12x + 4$

30. $20ab - 24a^2b^2 + 32a^3b^3$

Factoring, Regrouping: Binomial Square- the fourth type of factoring problems are called regrouping problems. There are two categories of regrouping problems, the binomial square type and the common factor type. These regrouping problems have 4 terms, and the **binomial square** type can be recognized because three of its four terms will be perfect squares.

To factor the binomial square regrouping problem:
  1. Regroup the first three terms.
  2. Factor the trinomial as a binomial square. The problem is now "the difference between two perfect squares" type factoring problem.
  3. Factor "the difference between two perfect squares" for the final factors.

Example

The problem:	$x^2 + 14x + 49 - a^2$
Regroup and factor:	$(x^2 + 14x + 49) - a^2$
Factor again:	$(x + 7)^2 - a^2$
The answer:	$(x + 7 - a)(x + 7 + a)$

Factor

1. $x^2 + 10x + 25 - y^2$

2. $x^2 - 6x + 9 - b^2$

3. $a^2 + 16a + 64 - r^2$

4. $a^2 - 2a + 1 - x^2$

5. $y^2 + 12y + 36 - z^2$

6. $25a^2 - 30ab + 9b^2 - 4$

7. $16b^2 + 56bc + 49c^2 - 9x^2$

8. $b^2 - 14b + 49 - 4x^2$

9. $c^2 - 4c + 4 - 25x^2$

10. $c^2 - 10c + 25 - 36b^2$

11. $z^2 - 18z + 81 - 49a^2$

12. $z^2 + 22z + 121 - 4y^2$

13. $9x^2 + 12x + 4 - a^2$

14. $25y^2 - 10y + 1 - c^2$

15. $4z^2 - 36z + 81 - b^2$

Factoring, Regrouping: Common Factors - The second type of regrouping problem, **common factors**, can be recognized because it has few if any perfect squares.

To factor the <u>common factor</u> regrouping problem:
1. Rearrange the terms if necessary, making sure the ratio of the first and second terms is the same as the ratio of the third and fourth terms.
2. Put parentheses around the 1st. and 2nd. terms and the 3rd. and 4th. terms.
3. Factor out a common factor from each set of parentheses, (even if it is only a 1).
4. Factor out a second common factor. (what is in both parentheses should be the same)

Example

The problem:	$x^2 - 3x + bx - 3b$
Regroup with parentheses:	$(x^2 - 3x) + (bx - 3b)$
Common factor, (x) and (b)	$x(x - 3) + b(x - 3)$
Common factor (x - 3)	$(x - 3)(x + b)$

Factor

1. $2xy - 6x + 3y - 9$

2. $x^2 + xy + 2x + 2y$

3. $x^2 - 3x + bx - 3b$

4. $ax + 3x + 4a + 12$

5. $ax + 5x + 6a + 30$

6. $2x^2 - 2xy + x - y$

7. $6x^2 + 2x + 6xy + 2y$

8. $4a^2 - ab + 4a - b$

9. $3a^2 - 6a + ab - 2b$

10. $2xy + 8y + x + 4$

11. $4a^2 + 4ab + 3a + 3b$

12. $3x^2 - 6x + xy - 2y$

13. $2x^2 - 3x + 4ax - 6a$

14. $3xy + 2x + 3y^2 + 2y$

15. $8ax - 2a + 4xy - y$

Factoring, Regrouping: Common Factors Continued - if the third term (-8a in the example) has a negative sign, the negative sign must be put inside the parentheses. The negative sign must then be factored out, which will change all the signs inside the parentheses and make it the same as the first parentheses. (See example)

Example

The problem:	$2a^2 - ab - 8a + 4b$
Regroup with parentheses:	$(2a^2 - ab) + (-8a + 4b)$
Common factor, (a) and (-4)	$a(2a - b) - 4(2a - b)$
Common factor $(2a - b)$	$(2a - b)(a - 4)$

Factor

1. $ax - 7a - 2x + 14$

2. $2x^2 + 2xy - 5x - 5y$

3. $ax + a - 2x - 2$

4. $3xy + y - 9x - 3$

5. $3a^2 - a - 6ab + 2b$

6. $2ax + x - 6a - 3$

7. $2ax - 3x - 4a + 6$

8. $3x^2 - x - 3xy + y$

9. $4ax - 8x - 3a + 6$

10. $3x^2 - 6x - xy + 2y$

11. $2a^2 + 3a - 8ax - 12x$

12. $4ab + b - 6a - 3$

13. $5a^2 - 15a - 3ax + 9x$

14. $xy - 3x - y^2 + 3y$

15. $7a^2 - ay - 7ab + by$

16. $3xy - 6x - y + 2$

17. $6x^2 - 4x - 3xy + 2y$

18. $4a^2 + 12a - ab - 3b$

Review of Factoring

These factoring problems will cover the following types:
1. Common factors
2. Difference between two perfect squares
3. Trinomials that factor as a binomial square
4. Regrouping - binomial square and common factors

Factor

1. $x^2 - 36$

2. $a^2 + 20a + 100$

3. $a^2 + 16a + 64 - r^2$

4. $2xy - 6x + 3y - 9$

5. $3x - 12$

6. $4a^3 + 3a^2 + 2a$

7. $z^2 + 16z + 64$

8. $b^2 + 14b + 49 - 4x^2$

9. $ax - 7a - 2x + 14$

10. $2a^2 + 3a - 8a - 12$

11. $x^2 - 16$

12. $7b + 8bc$

13. $4r - 10r^2$

14. $x^2 + xy + 2x + 2y$

15. $2x^2 + 2xy - 5x - 5y$

16. $12r - 16r^2 + 8r^3$

17. $16x^2 + 40x + 25$

18. $81c^2 + 36c + 4 - q^2$

19. $x^2 - 3x - bx + 3b$

20. $ax + a - 2x - 2$

Factoring, Trinomials - this type of trinomial does not have a common factor and it will have a <u>leading coefficient **greater** than 1</u>. Here are two ways to factor this type of problem:

1. The Regrouping Method
In this method, the middle term is replaced by two terms, whose sum equals the middle term. These two terms are listed as possible factors of the product of the first and third terms. This changes the problem to a "common factor regrouping" problem.

2. The Trial and Error Method.
In this method, experiments must by made with the first and last terms of the factors, to get products that equal the first and last terms of the trinomial. The "OI" portion of FOIL must then produce the middle term of the polynomial

Examples

$4x^2 + 12x + 5$		$4x^2 + 12x + 5$	
$4x^2 + 2x + 10x + 5$	Possible Factors	Possible factors	"OI"
$(4x^2 + 2x) + (10x + 5)$	$1 \bullet 20$	$(2x + 5)(2x + 1)$ *	$12x$ ←
$2x(2x + 1) + 5(2x + 1)$	$2 \bullet 10$ ←	$(4x + 5)(x + 1)$	$9x$
$(2x + 1)(2x + 5)$	$4 \bullet 5$	$(4x + 1)(x + 5)$	$21x$

Since the "OI" part of the first factors is $12x$, $(2x + 5)(2x + 1)$ are the factors.

Factor, using either method

1. $2x^2 + 3x + 1$

2. $2x^2 + 9x + 10$

3. $12x^2 + 13x + 3$

4. $3x^2 + 4x + 1$

5. $3x^2 + 5x + 2$

6. $2x^2 + 5x + 3$

7. $4x^2 + 5x + 1$

8. $4x^2 + 11x + 6$

9. $2x^2 + 9x + 9$

10. $4x^2 + 8x + 3$

11. $3x^2 + 10x + 3$

12. $8x^2 + 14x + 3$

Factoring, Trinomials Continued - if the third term of the trinomial is positive, the possible factors have the same signs, (both positive or both negative), and the signs are always the <u>same</u> as the sign of the <u>middle term</u>. In the following problems, the middle term is negative, thus all possible factors will be negative.

Examples

Regrouping Method   Trial and Error Method

$3x^2 - 13x + 4$ Possible Factors |Possible Factors   "OI"
$3x^2 - 12x - 1x + 4$ $-1 \bullet -12$ ← Add (-1) and (-12) to |$(3x - 2)(x - 2)$ $-8x$
$(3x^2 - 12x) + (-1x + 4)$ $-2 \bullet -6$ get (-13), the middle |$(3x - 4)(x - 1)$ $-7x$
$3x(x - 4) - 1(x - 4)$ $-3 \bullet -4$ term. |$(3x - 1)(x - 4)$ $\underline{-13x}$
$(x - 4)(3x - 1)$

Factor

1. $2x^2 - 7x + 6$ 6. $3x^2 - 7x + 2$ 11. $2x^2 - 9x + 10$

2. $2x^2 - 9x + 4$ 7. $12x^2 - 7x + 1$ 12. $4x^2 - 15x + 9$

3. $2x^2 - 7x + 5$ 8. $3x^2 - 13x + 4$ 13. $6x^2 - 5x + 1$

4. $4x^2 - 16x + 15$ 9. $2x^2 - 5x + 2$ 14. $6x^2 - 13x + 6$

5. $2x^2 - 5x + 3$ 10. $2x^2 - 13x + 20$ 15. $4x^2 - 21x + 5$

Factoring, Trinomials Continued - if the third term of the trinomial is negative, the possible factors must have **unlike** signs, and the largest of the two factors must take the sign of the middle term, (when using the regrouping method).

Examples

Regrouping Method

$4x^2 + 19x - 5$ Possible Factors
$4x^2 - 1x + 20x - 5$ $-1 \bullet +20$ ←
$(4x^2 - 1x) + (20x - 5)$ $-2 \bullet +10$
$x(4x - 1) + 5(4x - 1)$ $-4 \bullet +5$
$(4x - 1)(x + 5)$

Trial and Error Method

Possible Factors	"OI"
$(4x + 5)(x - 1)$	$+1x$
$(4x - 5)(x + 1)$	$-1x$
$(4x + 1)(x - 5)$	$-19x$
$(4x - 1)(x + 5)$ ←	$+19x$
$(2x - 5)(2x + 1)$	$-8x$
$(2x + 5)(2x - 1)$	$+8x$

Factor

1. $2x^2 + x - 1$
2. $2x^2 + x - 10$
3. $12x^2 + 5x - 3$
4. $3x^2 + 2x - 1$
5. $3x^2 + x - 2$

6. $2x^2 + 3x - 2$
7. $4x^2 + 3x - 1$
8. $8x^2 + 2x - 1$
9. $4x^2 + 5x - 6$
10. $2x^2 + 3x - 9$

11. $4x^2 + 4x - 15$
12. $3x^2 + 7x - 6$
13. $2x^2 + 5x - 3$
14. $8x^2 + 18x - 5$
15. $3x^2 + 8x - 3$

Factoring, Trinomials Continued - as stated on the previous page, if the third term of the trinomial is negative, the possible factors must have **unlike** signs, and the largest of the two factors must take the sign of the middle term.

Example (Regrouping Method)

$$2x^2 - 5x - 12$$
$$2x^2 - 8x + 3x - 12$$
$$(2x^2 - 8x) + (3x - 12)$$
$$2x(x - 4) + 3(x - 4)$$
$$(x - 4)(2x + 3)$$

Possible Factors
1 • −24
2 • −12
3 • −8 ←
4 • −6

Factor

1. $2x^2 - 11x - 6$

2. $2x^2 - 7x - 4$

3. $2x^2 - 3x - 5$

4. $4x^2 - 4x - 15$

5. $2x^2 - x - 3$

6. $3x^2 - 5x - 2$

7. $12x^2 - x - 1$

8. $3x^2 - 11x - 4$

9. $2x^2 - 3x - 2$

10. $4x^2 - 19x - 5$

11. $2x^2 - 3x - 20$

12. $3x^2 - 8x - 3$

13. $2x^2 - x - 10$

14. $4x^2 - 9x - 9$

15. $6x^2 - x - 1$

Review of all the types of factoring problems covered thus far.

1. $10x - 15$

2. $x^2 - 49$

3. $x^2 + 16x + 64$

4. $x^2 - 6x + 9 - b^2$

5. $5x^2 - 10x + 3ax - 6a$

6. $5x^2 - x - 4$

7. $9x^2 - 16$

8. $15x^2 - 10x$

9. $x^2 + 10x + 25 - 9y^2$

10. $9x^2 + 6x + 1$

11. $3x^2 + 5x - 2$

12. $by + 5b + y^2 + 5y$

13. $a^2 - 4a + 4 - 25b^2$

14. $25x^2 - 36$

15. $9x^2 + 12x + 4$

16. $3bc + b$

17. $4a^2 + 9a + 5$

18. $4a^2 - ab - 12a + 3b$

19. $4ax - 6x + 2a^2 - 3a$

20. $2a^3 - 8a^2 - 12a$

21. $81a^2 + 36a + 4 - r^2$

22. $4x^2 - 21x + 5$

23. $25a^6 - 16$

24. $36x^2 + 12x + 1$

Factoring, Trinomials Continued - whereas the trinomials thus far have all had leading coefficients that were <u>larger than 1</u>, the next group of polynomials have a leading coefficient <u>of 1</u>. These problems are <u>less difficult</u> than the previous trinomials.

The Regrouping Method
List the factors of the third term, put the correct signs with the factors, and pair each factor with "x" inside a parentheses.

The Trial and Error Method
To find the factors, ask this question: "What numbers multiply to give you the last term and add to give you the middle term?" Pair each number with "x" inside a parentheses.

Examples

$x^2 + 7x + 10$
$(x+2)(x+5)$

Possible Factors
+1 • +10
+2 • +5 ←

$x^2 + 7x + 10$

Possible Factors	"OI"
$(x+10)(x+1)$	$11x$
$(x+2)(x+5)$	$7x$ ←

Factor

1. $x^2 + 5x + 6$

2. $x^2 + 6x + 8$

3. $x^2 + 8x + 12$

4. $x^2 + 4x + 3$

5. $x^2 + 10x + 16$

6. $x^2 + 7x + 12$

7. $x^2 + 11x + 28$

8. $x^2 + 14x + 45$

9. $x^2 + 15x + 54$

10. $x^2 + 16x + 63$

11. $x^2 + 17x + 72$

12. $x^2 + 19x + 90$

13. $x^2 + 8x + 7$

14. $x^2 + 11x + 18$

15. $x^2 + 11x + 24$

16. $x^2 + 12x + 27$

17. $x^2 + 14x + 40$

18. $x^2 + 11x + 30$

Factoring, Trinomials Continued - since the third term of these trinomials is positive, these trinomials have factors that have the have **the same sign** as the middle term.

Examples

Regrouping Method   Trial and Error Method

$x^2 - 10x + 21$

Regrouping	Possible Factors		Trial and Error	Possible Factors	"OI"
$x^2 - 10x + 21$	$-1 \cdot -21$		$x^2 - 10x + 21$	$(x-21)(x-1)$	$-22x$
$(x-3)(x-7)$	$-3 \cdot -7$ ←			$(x-3)(x-7)$	$-10x$ ←

Factor

1. $x^2 - 3x + 2$
2. $x^2 - 13x + 36$
3. $x^2 - 16x + 60$
4. $x^2 - 13x + 30$
5. $x^2 - 13x + 40$
6. $x^2 - 8x + 15$
7. $x^2 - 20x + 99$
8. $x^2 - 7x + 6$

9. $x^2 - 15x + 44$
10. $x^2 - 8x + 12$
11. $x^2 - 12x + 11$
12. $x^2 - 12x + 32$
13. $x^2 - 13x + 22$
14. $x^2 - 16x + 15$
15. $x^2 - 9x + 8$
16. $x^2 - 10x + 16$

17. $x^2 - 14x + 33$
18. $x^2 - 11x + 10$
19. $x^2 - 17x + 66$
20. $x^2 - 12x + 20$
21. $x^2 - 15x + 54$
22. $x^2 - 15x + 50$
23. $x^2 - 16x + 55$
24. $x^2 - 9x + 20$

Factoring, Trinomials Continued - since the third term of these trinomials is <u>negative</u>, these trinomials have factors that have unlike signs, and the larger number will take the sign of the middle term, (positive in these problems).

Examples

Regrouping Method   Trial and Error Method
                                       $x^2 + 2x - 15$

	Possible Factors		Possible factors	"OI"
$x^2 + 2x - 15$	$-1 \bullet +15$		$(x + 15)(x - 1)$	$14x$
$(x + 5)(x - 3)$	$-3 \bullet +5$ ←		$(x - 15)(x - 1)$	$-14x$
			$(x + 5)(x - 3)$ *	$+2x$ ←
			$(x - 5)(x + 3)$	$-2x$

Factor

1. $x^2 + 4x - 21$
2. $x^2 + 4x - 5$
3. $x^2 + 6x - 16$
4. $x^2 + 8x - 20$
5. $x^2 + 3x - 18$
6. $x^2 + x - 6$
7. $x^2 + 8x - 33$

8. $x^2 + 2x - 3$
9. $x^2 + 7x - 30$
10. $x^2 + 2x - 80$
11. $x^2 + 4x - 77$
12. $x^2 + 4x - 12$
13. $x^2 + 5x - 24$
14. $x^2 + x - 30$

15. $x^2 + x - 2$
16. $x^2 + 2x - 15$
17. $x^2 + 6x - 40$
18. $x^2 + 2x - 8$
19. $x^2 + 5x - 6$
20. $x^2 + 7x - 18$
21. $x^2 + 5x - 36$

Factoring, Trinomials Continued - since the third term of these trinomials is also negative, the factors of these trinomials will have unlike signs, with the larger number taking the sign of the middle term, (negative in these problems).

Examples

Regrouping Method

$x^2 - 3x - 18$ Possible Factors
$-1 \cdot -18$
$(x - 6)(x + 3)$ $+2 \cdot -9$
$+3 \cdot -6$ ←

Trial and Error Method

$x^2 - 3x - 18$ Possible factors "OI"
$(x + 1)(x - 18)$ $-17x$
$(x - 1)(x + 18)$ $+17x$
$(x + 2)(x - 9)$ $-7x$
$(x - 2)(x + 9)$ $+7x$
$(x + 3)(x - 6)$ $-3x$
$(x - 3)(x + 6)$ * $+3x$ ←

---

1. $x^2 - x - 20$
2. $x^2 - 3x - 28$
3. $x^2 - 2x - 48$
4. $x^2 - 7x - 44$
5. $x^2 - x - 12$
6. $x^2 - 2x - 99$
7. $x^2 - 8x - 33$
8. $x^2 - 5x - 50$

9. $x^2 - x - 56$
10. $x^2 - 9x - 22$
11. $x^2 - 6x - 55$
12. $x^2 - 3x - 88$
13. $x^2 - 8x - 9$
14. $x^2 - 4x - 5$
15. $x^2 - 5x - 24$
16. $x^2 - 4x - 60$

17. $x^2 - 5x - 14$
18. $x^2 - 2x - 8$
19. $x^2 - x - 72$
20. $x^2 - 2x - 35$
21. $x^2 - 7x - 8$
22. $x^2 - 5x - 36$
23. $x^2 - 8x - 20$
24. $x^2 - 6x - 40$

Review of Trinomials

1. $x^2 - 8x - 9$

2. $3x^2 - 8x + 5$

3. $x^2 + 8x + 7$

4. $4x^2 - 3x - 1$

5. $x^2 + 12x + 11$

6. $x^2 - 11x + 18$

7. $x^2 + 7x - 18$

8. $2x^2 + 5x - 7$

9. $2x^2 - 9x + 7$

10. $4x^2 + 12x + 5$

11. $x^2 + 5x - 50$

12. $x^2 - 5x - 14$

13. $5x^2 - x - 6$

14. $x^2 + 7x + 10$

15. $5x^2 - 11x + 2$

16. $x^2 - x - 42$

17. $x^2 + 4x - 32$

18. $6x^2 - 19x - 7$

19. $x^2 - 12x + 27$

20. $4x^2 + x - 3$

21. $3x^2 + 11x + 6$

Review of Factoring

1. $x^2 + 10x + 25$
2. $a^2 + ab + 4a + 4b$
3. $x^2 - 16$
4. $x^2 - 12x + 36 - b^2$
5. $3r - 9$
6. $2x^2 - 7x + 6$
7. $x^2 - x - 30$
8. $x^2 - 14x + 49$
9. $2x^2 + 9x + 9$
10. $2x^2 - 3x - 20$
11. $11xb - 9b$
12. $x^2 + 3x - 18$
13. $4x^2 + 4x - 15$
14. $3x^2 - 7x - 10$
15. $ax - 2x + a - 2$
16. $1 - a^2b^2$
17. $x^2 - 13x + 30$
18. $a^2 + 2a + 1$
19. $2a^2 + 7a + 3$
20. $4x^2 - 9$
21. $3x^2 - 6x + xy - 2y$
22. $10r^2x^2 + 25rx$
23. $9a^2 - 48a + 64 - 25b^2$
24. $4y^2 - 13y + 3$
25. $b^2 + 14b + 49 - y^2$
26. $36a^2 - 9b^2$
27. $4ax - 8x - 3a + 6$

Factoring, Combination Problems - The last type of factoring problems are problems that can be factored more than once. They include a combination of different type factoring techniques.

The first type are problems that require you to twice factor the "difference between two perfect squares". Do not attempt to factor the sum of two perfect squares. $(x^2 + 1)$

$$\begin{array}{c} \textbf{Example} \\ x^4 - 1 \\ (x^2 - 1)(x^2 + 1) \\ (x - 1)(x + 1)(x^2 + 1) \end{array}$$

Factor

1. $x^4 - 1$

2. $1 - y^4$

3. $x^4 - y^4$

4. $x^4 - 16$

5. $a^4 - 16$

6. $16 - x^4$

7. $x^4 - 81$

8. $y^4 - 81$

9. $a^4 - 16b^4$

10. $81 - x^4$

11. $81b^4 - c^4$

12. $81r^4 - 1$

13. $16c^4 - 81y^4$

14. $81b^4 - 16$

15. $x^4 - 256$

Factoring, Combination Problems Continued - in these problems the order must be changed. Take out a common factor if need be, then factor the difference between two perfect squares.

Examples

$$-x^2 + 4$$
$$4 - x^2$$
$$(2 - x)(2 + x)$$

$$-18 + 2x^2$$
$$2x^2 - 18$$
$$2(x^2 - 9)$$
$$2(x - 3)(x + 3)$$

Factor

1. $-x^2 + 81$

2. $-y^2 + 25$

3. $-a^2 + 16$

4. $-36 + x^2$

5. $-9 + x^2$

6. $-32 + 2x^2$

7. $-20x + 5x^3$

8. $-2 + 8x^2$

9. $-3x + 3x^3$

10. $-48 + 3x^2$

11. $-4 + x^2$

12. $-25 + y^2$

13. $-144 + 9x^2$

14. $-72 + 50x^2$

15. $-2 + 50y^2$

16. $-81 + x^2$

17. $-8 + 2x^2$

18. $-27 + 3x^2$

19. $-98x + 2x^3$

20. $-108x + 3x^3$

21. $-49 + x^2$

22. $-1 + a^2$

23. $-64 + b^2$

Factoring Combination Problems Continued - these two types of problems have a common factor. Take out the common factor, then factor what is left.

Examples

a. $2x^2 - 8$
$2(x^2 - 4)$
$2(x-2)(x+2)$

b. $2x^2 + 18x + 40$
$2(x^2 + 9x + 20)$
$2(x+5)(x+4)$

Factor

1. $2x^2 - 18$

2. $2x^2 - 200$

3. $3x^2 - 75$

4. $2x^2 + 20x + 42$

5. $2x^2 + 14x - 36$

6. $3x^2 + 15x - 108$

7. $3x^2 - 432$

8. $4x^2 - 144$

9. $2x^2 + 26x + 60$

10. $3x^2 - 3x - 60$

11. $5x^2 - 245$

12. $3x^2 - 12$

13. $3x^3 + 9x^2 - 30x$

14. $3x^3 - 27x$

15. $6x^2 + 54x + 48$

16. $5x^3 - 20x$

17. $2x^3 - 2x^2 - 24x$

18. $2x^3 - 2x$

Factoring Review

1. $2x^2 + 14x - 36$

2. $3a^2 - a - 6ac + 2c$

3. $2x^2 - 9x + 10$

4. $6x + 10x^2$

5. $x^2 + 24x + 23$

6. $x^2 - 8x - 9$

7. $36y^2 + 12y + 1$

8. $-64 + b^2$

9. $4y^2 - 3ay + 8y - 6a$

10. $7xy - 6y$

11. $12x^2 + 11x + 2$

12. $25y^2 - 10y + 1 - 9a^2$

13. $2x^2 - 5x - 12$

14. $2x^2 - 100$

15. $12r - 16r^2 + 6r^3$

16. $x^2 - 9x + 14$

17. $25x^2 - 9$

18. $x^4 - 81$

19. $8 - 18a$

20. $6x^2 + 7x - 3$

21. $x^2 - x - 30$

22. $2x^2 - 18$

23. $6x^2 + x - 1$

24. $2y^2 + 20y + 42$

Cumulative Review 6

Solve, put on the number line, and give the solution set:

1. $6 \geq x + 10$

   ―――――|―――
         0
   x = {   }

2. $-2 < x + 4 < 3$

   ―――――|―――
         0
   x = {   }

Solve:

3. $8x + 7 - 3x = 12 - x - 29$

4. $\dfrac{2x}{5} + 10 = 18$

Simplify:

5. 4^{-3}

6.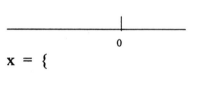

Multiply:

7. $(2a^2c^4)^3(3a^3c)^2$

8. $\begin{array}{r} 5x - 7 \\ \underline{3x + 2} \end{array}$

Name the property:

9. $a = 13 - 8$
   $13 - 8 = 5$
   $a = 5$ _____

10. $\tfrac{1}{2}(2x) = \tfrac{1}{2}(24)$ _____

11. $x = 4(3y)$
    $x = 12y$ _____

Simplify:

12. $17 - (2x + 8)$

13. $3(7x - 10)$

14. $9 + (-6 - 12)$

15. $(-3)(-5)(-4) - (-2)(-1)(-6)$

Chapter 9
Quadratic Equations and Rational Expressions

Solving Quadratic Equations by Factoring - a <u>quadratic</u> equation has <u>one</u> varable, and one of the exponents of that variable must be a "2", and no exponent can be larger than 2. $3x^2 - 7x + 2 = 0$ is a quadratic equation. (one variable{x}, exponent of 2)

If two terms are multiplied and their product is "0", then one of the two terms must be equal to 0. If those terms are the factors of a quadratic equation, the solutions of the quadratic equation are called the <u>roots of the equation</u>.

Examples

a. $(x)(y) = 0$
   $x = 0$ or $y = 0$

b. $2x(x - 3) = 0$
   $2x = 0$ or $x - 3 = 0$

   $\dfrac{2x}{2} = \dfrac{0}{2}$ $x - 3 = 0$
    $+3\ +3$
   $x = 0$ or $x = +3$

c. $(x + 5)(x - 7) = 0$
   $x + 5 = 0$ or $x - 7 = 0$
   $-5\ -5+7\ +7$
   $x = -5$ or $x = +7$
   [note the signs of the roots are opposite those of the factors]

Solve for the variables

1. $(a)(b) = 0$

2. $(p)(q) = 0$

3. $x(x - 4) = 0$

4. $3a(a + 6) = 0$

5. $(x - 1)(x + 10) = 0$

6. $(y + 6)(y - 11) = 0$

7. $(r)(t) = 0$

8. $2b(b - 9) = 0$

9. $(x + 2)(x + 5) = 0$

10. $6x(x - 7) = 0$

11. $(y - 4)(y - 1) = 0$

12. $a(a + 9) = 0$

13. $5x(x - 12) = 0$

115

More Solving Quadratic Equations by Factoring - if the quadratic equation has not been factored, then it <u>must be factored</u>.

Before a quadratic equation can be factored, it must be equal to "0". If it is not equal to 0, whatever is on the <u>right side of the equation</u> **must be moved** to the left side of the equation.

Examples

a. $(3x+2)(2x-7) = 0$
$3x+2 = 0$ or $2x-7 = 0$
$\quad -2 \;\; -2 \qquad\quad +7 \;\; +7$
$\quad \dfrac{3x}{3} = \dfrac{-2}{3} \qquad \dfrac{2x}{2} = \dfrac{7}{2}$
$\quad x = -\dfrac{2}{3} \qquad\quad x = \dfrac{7}{2}$

b. $2x^2 = 6x$
$\quad -6x \;\; -6x$
$2x^2 - 6x = 0$
$2x(x-3) = 0$
$2x = 0$ or $x-3 = 0$
$x = 0$ or $x = +3$

c. $4x^2 - 9 = 0$
$(2x-3)(2x+3) = 0$
$2x-3 = 0$ or $2x+3 = 0$
$\quad +3 \;\; +3 \qquad\quad -3 \;\; -3$
$\quad \dfrac{2x}{2} = \dfrac{3}{2} \qquad \dfrac{2x}{2} = \dfrac{-3}{2}$
$\quad x = \dfrac{3}{2}$ or $x = -\dfrac{3}{2}$

Solve the quadratic equations; factor if necessary

1. $(3x+5)(6x+1) = 0$

2. $(4x-3)(5x-4) = 0$

3. $3x^2 = -9x$

4. $5x^2 = 20x$

5. $16x^2 - 9 = 0$

6. $4x^2 - 25 = 0$

7. $(10x-5)(4x+3) = 0$

8. $x^2 = +9$

9. $(4x+5)(4x+3) = 0$

10. $6x^2 = -2x$

11. $9x^2 - 4 = 0$

12. $9x^2 - 3x = 0$

More Solving Quadratic Equations by Factoring - use the rules for factoring trinomials to solve these quadratic equations.

Examples

a. $x^2 + 9x - 22 = 0$
   $(x + 11)(x - 2) = 0$
   $x + 11 = 0$ or $x - 2 = 0$
   $\underline{-11 \quad -11} \quad \underline{+2 \quad +2}$
   $x \quad = -11$ or $x \quad = +2$

b. $x^2 - 9x \quad = -20$
   $\underline{\quad\quad\quad +20 \quad +20}$
   $x^2 - 9x + 20 = \quad 0$
   $(x - 4)(x - 5) = \quad 0$
   $x = +4$ or $x = +5$

c. $6x^2 + 5x - 6 = 0$
   $(6x^2 - 4x) + (9x - 6) = 0$
   $2x(3x - 2) + 3(3x - 2) = 0$
   $(3x - 2)(2x + 3) = 0$
   $x = \frac{2}{3}$ or $x = -\frac{3}{2}$

Find the roots of the quadratic equations by factoring

1. $x^2 + 7x + 12 = 0$

2. $x^2 + 2x - 15 = 0$

3. $x^2 - 8x = -7$

4. $x^2 + 12x = -35$

5. $3x^2 + 5x + 2 = 0$

6. $4x^2 - 15x + 9 = 0$

7. $x^2 - 3x - 18 = 0$

8. $2x^2 + 3x = -1$

9. $x^2 - x = +20$

10. $3x^2 + 7x - 6 = 0$

11. $x^2 + 13x = -12$

12. $6x^2 - 13x - 5 = 0$

13. $x^2 - 6x = -9$

14. $x^2 - 11x + 10 = 0$

15. $3x^2 + 4x - 7 = 0$

Simplifying Rational Expressions by Cancelling - rational expressions are algebraic expressions in fraction form. To simplify rational expressions, identical binomials can cancel out each other, as long as one is a numerator and the other a denominator. Any part of a monomial can cancel with any part of another monomial.

Examples

a. $\dfrac{(x+2)(x-1)}{(x+2)} = \dfrac{\cancel{(x+2)}(x-1)}{\cancel{(x+2)}} = x - 1$ b. $\dfrac{(2x+5)4}{2(x-4)(2x+5)} = \dfrac{\cancel{(2x+5)}\cancel{4}^2}{\cancel{2}(x-4)\cancel{(2x+5)}} = \dfrac{2}{x-4}$

Simplify

1. $\dfrac{(x+6)(x+1)}{(x+6)}$

2. $\dfrac{(2x+1)(x-3)}{x-3}$

3. $\dfrac{x-5}{(x-5)(x+5)}$

4. $\dfrac{(4x+5)(x-1)}{6(4x+5)}$

5. $\dfrac{(x+3)(x-2)}{(x+5)(x+3)}$

6. $\dfrac{2(x+6)}{2(x-5)(x+6)}$

7. $\dfrac{6(x+1)}{3(x+2)(x+1)}$

8. $\dfrac{15(x-4)}{(x-4)(x-3)3}$

9. $\dfrac{4(4x+3)(3x-7)}{2(3x-7)}$

10. $\dfrac{(x-5)(x+2)}{(x+2)(x-3)(x-5)}$

11. $\dfrac{9(x+7)(2x-1)}{3(3x-1)(2x-1)(x+7)}$

12. $\dfrac{(3x-4)(x+6)5}{5(3x-4)}$

13. $\dfrac{7(x-9)(2x+5)}{(x-9)(4x-1)(2x+5)21}$

14. $\dfrac{(x-6)(x-10)}{(x-6)}$

15. $\dfrac{2x+3}{3(x-2)(2x+3)}$

16. $\dfrac{(x+9)(x+9)}{x+9}$

17. $\dfrac{3(x+2)(3x+4)}{3(x+2)}$

18. $\dfrac{(2x+1)(3x-4)}{2(3x-4)}$

19. $\dfrac{12(x+2)(x-6)}{(x-6)(x+2)2}$

20. $\dfrac{20(x-3)}{(x-3)(x+1)4}$

21. $\dfrac{(a+2)(a-3)}{(a-3)(a+6)}$

More Simplifying Rational Expressions - if either the numerator of denominator can be factored, do so, then cancel any monomials or <u>identical binomials</u>.

Examples

a. $\dfrac{x^2-16}{x-4} = \dfrac{\cancel{(x-4)}(x+4)}{\cancel{x-4}} = x+4$

b. $\dfrac{6x^2-30x}{(x-5)^2} = \dfrac{6x\cancel{(x-5)}}{\cancel{(x-5)}(x-5)} = \dfrac{6x}{x-5}$

Factor if necessary, then simplify

1. $\dfrac{x^2-25}{x-5}$

2. $\dfrac{x^2-81}{x+9}$

3. $\dfrac{2x^2-14x}{x^2-49}$

4. $\dfrac{(x+8)^2}{x^2-64}$

5. $\dfrac{(x-5)^2}{3x-15}$

6. $\dfrac{(4x-1)^2}{8x-2}$

7. $\dfrac{3x+2}{(3x+2)^2}$

8. $\dfrac{5a^2+25a}{a^2-25}$

9. $\dfrac{6x^2-9x}{4x^2-9}$

10. $\dfrac{x^2-49}{(x+7)^2}$

11. $\dfrac{x^2-y^2}{x-y}$

12. $\dfrac{16a^2-b^2}{4a+b}$

13. $\dfrac{10x^2-80x}{x^2-64}$

14. $\dfrac{9x^2-4}{3x^2+2x}$

15. $\dfrac{4x^2-25y^2}{2x-5y}$

16. $\dfrac{(2x-y)^2}{2xy-y^2}$

17. $\dfrac{x^2-64}{(x+8)^2}$

18. $\dfrac{8x^2-32x}{x^2-16}$

More Simplifying Rational Expressions - before simplifying rational expressions by factoring and cancelling, check the polynomials to make sure they are in the <u>same order</u>, usually the highest power of the variable to the lowest power of the variable. If they are not in the same order, change them, then proceed to factor and simplify.

Examples

a. $\dfrac{x^2-36+5x}{27x+3x^2} = \dfrac{x^2+5x-36}{3x^2+27x} = \dfrac{\cancel{(x+9)}(x-4)}{3x\cancel{(x+9)}} = \dfrac{(x-4)}{3x}$

b. $\dfrac{3x^2+4+13x}{-4+7x+2x^2} = \dfrac{3x^2+13x+4}{2x^2+7x-4} = \dfrac{(3x+1)\cancel{(x+4)}}{(2x-1)\cancel{(x+4)}} = \dfrac{3x+1}{2x-1}$

Put in proper order if needed, then factor and simplify

1. $\dfrac{x^2-2x-3}{3+4x+x^2}$

2. $\dfrac{x^2-5+4x}{2x+10}$

3. $\dfrac{x^2-3x-10}{-4x-12+x^2}$

4. $\dfrac{x^2-x-6}{-15+5x}$

5. $\dfrac{3x^2+2-5x}{x^2-1}$

6. $\dfrac{2-7x+3x^2}{x^2-7x+10}$

7. $\dfrac{12x+9}{4x^2+3+7x}$

8. $\dfrac{x^2+7x-18}{x^2-4}$

9. $\dfrac{-8x+x^2-20}{-7x+x^2-30}$

10. $\dfrac{14x+2x^2}{x^2-49}$

11. $\dfrac{x^2+3x-18}{-9+x^2}$

12. $\dfrac{2x^2-x-15}{2x^2+7x+5}$

13. $\dfrac{6x-7+x^2}{3x-4+x^2}$

14. $\dfrac{4x+x^2}{x^2+9x+20}$

15. $\dfrac{6x^2-2+x}{6x^2+1-5x}$

16. $\dfrac{81-18x+x^2}{(x-9)^2}$

Simplifying Rational Exprssions by Multiplication - in multiplication, when the numerators and denominators are <u>monomials</u>, parts of the numerators can be cancelled with parts of the denominator. Two of the ways to work this type of problem are:

Examples

a. Multiplication then simplify: $\dfrac{6a^2}{5ay} \bullet \dfrac{2y^2}{3a} = \dfrac{12a^2y^2}{15a^2y} = \dfrac{12y}{15} = \dfrac{4y}{5}$

b. Factor then cancel: $\dfrac{6a^2}{5ay} \bullet \dfrac{2y^2}{3a} = \dfrac{2\bullet\cancel{3}\bullet\cancel{a}\bullet\cancel{a}\bullet 2\bullet\cancel{y}\bullet y}{5\bullet\cancel{a}\bullet\cancel{y}\bullet\cancel{3}\bullet\cancel{a}} = \dfrac{2\bullet 2\bullet y}{5} = \dfrac{4y}{5}$

Simplify

1. $\dfrac{7x}{14y} \bullet \dfrac{6y^2}{x}$

2. $\dfrac{6a}{21b^2} \bullet \dfrac{7b}{3}$

3. $\dfrac{3xy}{4yz} \bullet \dfrac{6x^2}{3y^2}$

4. $\dfrac{-14a^2}{18z^2} \bullet \dfrac{4xz}{21a}$

5. $\dfrac{4a^3xy}{12bc} \bullet \dfrac{24b^2c}{6ayx}$

6. $\dfrac{9ayx}{3xyz} \bullet \dfrac{7z}{-18a^2}$

7. $\dfrac{10n^3y}{6x^2} \bullet \dfrac{12n^2x^2}{25n^4y^2}$

8. $\dfrac{6a^2n}{8n^2} \bullet \dfrac{12n}{9a^3}$

9. $\dfrac{(xy)^2}{z} \bullet \dfrac{za^2}{ax^2y}$

10. $\dfrac{6x^3y}{9z^2} \bullet \dfrac{4z^2x}{10y^3z}$

11. $\dfrac{7ab^2}{11c^3} \bullet \dfrac{44c^2}{21a^2b}$

12. $\dfrac{3x^2y^4}{14xz^4} \bullet \dfrac{7z^3}{12xy^2}$

More Multiplication of Rational Expressions - when multiplying rational expressions containing binomials and trinomials, factor them, then cancel <u>identical binomials</u> and **parts** of any <u>monomials</u> if possible.

Examples

a. $\dfrac{5x-5}{3} \times \dfrac{9}{x-1} = \dfrac{5(x-1)}{3} \times \dfrac{9^3}{x-1} = \dfrac{5 \cdot 3}{1} = 15$

b. $\dfrac{x^2-2x}{x^2-10x+25} \cdot \dfrac{x^2-25}{x^2+3x-10} = \dfrac{x(x-2)}{(x-5)(x-5)} \cdot \dfrac{(x-5)(x+5)}{(x+5)(x-2)} = \dfrac{x}{x-5}$

Multiply

1. $\dfrac{3x-6}{4} \times \dfrac{8}{x-2}$

2. $\dfrac{6}{2x-8} \cdot \dfrac{x-4}{x}$

3. $\dfrac{b}{b+3} \times \dfrac{b^2-9}{b^2}$

4. $\dfrac{3a^2+9a}{4} \cdot \dfrac{2}{3a}$

5. $\dfrac{x^2-15x+50}{x^2-9x+20} \cdot \dfrac{x^2-11x+24}{x^2-18x+80}$

6. $\dfrac{a^2-2a}{a^2+2a-8} \times \dfrac{a^2+a-12}{2a^3-6a^2}$

7. $\dfrac{x^2-16}{3x^2-13x+4} \times \dfrac{3x^2+2x-1}{2x+8}$

8. $\dfrac{x^2+8x+12}{x^2-7x+12} \cdot \dfrac{x^2+x-20}{x^2+11x+30}$

Multiplication of Rational Expressions Continued

9. $\dfrac{x^2+7x+10}{3} \bullet \dfrac{6x}{x+2}$

10. $\dfrac{3x+15}{x^2+x-12} \times \dfrac{x-3}{x+5}$

11. $\dfrac{3x-6}{x^2-9} \times \dfrac{x+3}{x^2-2x}$

12. $\dfrac{3xy-3x}{x} \bullet \dfrac{2}{y^2-1}$

13. $\dfrac{b^2}{3b} \times \dfrac{x+6}{x^2+7x+6}$

14. $\dfrac{x^2-x-6}{3x^2+10x+8} \bullet \dfrac{3x^2-5x-12}{x^2-2x-3}$

15. $\dfrac{x^2-25}{-15+2x+x^2} \times \dfrac{-6-x+x^2}{2x^2-9x-5}$

16. $\dfrac{4x^2+5x-6}{2x+2} \bullet \dfrac{x^2-1}{4x^2-7x+3}$

17. $\dfrac{x^2+5x+6}{x^2+x-2} \times \dfrac{x^2+2x-3}{x^2+7x+12}$

18. $\dfrac{x^2+2x-3}{2x^2+3x-5} \bullet \dfrac{2x^2-x-15}{x^2-9}$

Simplifying Rational Expressions by Division - use the same steps used in multiplication **except** in division <u>invert (turn upside down) the divisor</u>. (the fraction to the right of the division sign)

Examples

a. $\dfrac{x^2y}{4xy^2} \div \dfrac{8x^2y^2}{16x^2y}$

$= \dfrac{x^2y}{4xy^2} \times \dfrac{16x^2y}{8x^2y^2} = \dfrac{16x^4y^2}{32x^3y^4} = \dfrac{x}{2y^2}$

b. $\dfrac{x^2+4x}{2} \div (x+4)$

$= \dfrac{x(x+4)}{2} \bullet \dfrac{1}{(x+4)} = \dfrac{x}{2}$

Simplify

1. $\dfrac{x}{3x} \div \dfrac{5x^4}{3y^3}$

2. $\dfrac{x^2y}{2z} \div \dfrac{x^2y^2}{z^2}$

3. $\dfrac{13x^2}{14z} \div \dfrac{26x^3}{70z^2}$

4. $\dfrac{x^2-36}{2a} \div (x+6)$

5. $\dfrac{x^2-81}{5x} \div (x-9)$

6. $\dfrac{x^2-x-20}{x+5} \div \dfrac{x+4}{1}$

7. $(x+7) \div \dfrac{x^2-49}{3}$

8. $(x+5) \div \dfrac{x^2+6x+5}{2}$

9. $\dfrac{a}{a^2-25} \div \dfrac{a^2}{a+5}$

10. $\dfrac{a}{x^2+3x+2} \div \dfrac{a^2}{x+2}$

Division of Rational Expressions Continued - first factor all binomials and trinomials

1. $\dfrac{x^2+x-2}{x^2+5x+6} \div \dfrac{x^2+2x-3}{x^2+7x+12}$

2. $\dfrac{x^2+2x-15}{x^2-x-30} \div \dfrac{x^2-3x-18}{x^2-2x-24}$

3. $\dfrac{2x^2+7x-15}{x+5} \div \dfrac{9x^2-4}{3x+2}$

4. $\dfrac{x^2+3x-10}{x^2+8x+15} \div \dfrac{x^2-6x+8}{x^2-x-12}$

5. $\dfrac{x^2-x-6}{2x^2-7x-15} \div \dfrac{x^2+6x+8}{2x^2+11x+12}$

6. $\dfrac{6+3x}{x^2+5x+6} \div \dfrac{8+4x}{x^2+2x-3}$

7. $\dfrac{x^2+8x+16}{y^2-6y+9} \div \dfrac{2x+8}{3y-9}$

8. $\dfrac{x^2-2x-8}{x+5} \div \dfrac{3x^2-10x-8}{3x+2}$

9. $\dfrac{x^2+x-2}{x^2+8x+12} \div \dfrac{x^2+2x-3}{x^2-2x-15}$

10. $\dfrac{3x^2-5x-2}{x^2-6x+8} \div \dfrac{9x^2+6x+1}{6x+2}$

Simplifying Rational Expressions Containing Multiplication and Division - when the rational expression contains multiplication and division, invert only the fraction immediately <u>following the division sign</u>.

Example

a. $\dfrac{x}{x+3} \div \dfrac{3x^2}{3x+9} \times \dfrac{x^2+4x+3}{x^2-9}$

$= \dfrac{x}{x+3} \times \dfrac{3(x+3)}{3x^2} \times \dfrac{(x+3)(x+1)}{(x+3)(x-3)} = \dfrac{x+1}{x(x-3)}$

Simplify

1. $\dfrac{y+3}{y+2} \times \dfrac{3}{y+3} \div \dfrac{6}{y+2}$

2. $\dfrac{x^2}{4x} \div \dfrac{x^2y^2-4}{2x} \cdot \dfrac{xy-2}{xy}$

3. $\dfrac{x^2-4}{x^2} \div \dfrac{x^2-5x+6}{x^2-3x} \cdot \dfrac{4x}{x^2+6x+8}$

4. $\dfrac{3x^2+6x}{x^2-4} \div \dfrac{x-6}{x-2} \times \dfrac{x-6}{6x^2}$

5. $\dfrac{x-3}{x^2+2x} \cdot \dfrac{x^2+5x}{3x-9} \div \dfrac{x^3+5x^2}{6x+12}$

6. $\dfrac{2x^2+3x+1}{x^2-x-6} \cdot \dfrac{x^2-4}{2x^2-x-1} \div \dfrac{x^2-x-2}{x^2-2x-15}$

7. $\dfrac{(x+2)^2}{3x^2+12x} \div \dfrac{x+2}{15x^2} \times \dfrac{5x-15}{x^2+x-12}$

8. $\dfrac{3x^2+8x+4}{x^2-16} \div \dfrac{3x^2-x-2}{x^2+10x+24} \cdot \dfrac{x^2-6x+8}{x^2+8x+12}$

Chapter 9 Review

Solve for the variable

1. $(x-1)(x+9) = 0$

2. $4x^2 - 12x = 0$

3. $7x^2 - 8x + 1 = 0$

4. $x^2 + 7x = -12$

5. $(x)(y) = 0$

6. $9x^2 - 25 = 0$

7. $4x(2x+3) = 0$

Simplify

8. $\dfrac{(x+3)(x-7)}{3(x-7)}$

9. $\dfrac{x^2-16}{x^2-3x-4} \times \dfrac{3x^2+2x-1}{2x+8}$

10. $\dfrac{x^2-10x+21}{-9x+3x^2}$

11. $\dfrac{a^2-a-12}{4a} \div \dfrac{a-4}{8}$

12. $\dfrac{x^2-5x-6}{x^2+8x+15} \bullet \dfrac{4x+20}{x^2-7x+6} \div \dfrac{x+1}{x^2+6x-7}$

13. $\dfrac{10x^2}{6z^3} \times \dfrac{12x^2z^4}{25x^2z}$

Cumulative Review 7

Factor:

1. $8x + 6x^2$

2. $x^2 - 3x - 18$

3. $-25 + x^2$

4. $8x^2 + 10x + 3$

5. $x^2 - 12x + 36 - 9a^2$

Solve and give the solution set:

6. $4x + 9 \leq 2x + 19$

   $x = \{ \qquad \}$

Solve:

7. $7x - 31 = 56$

8. $|2x + 1| = 9$

9. Simplify: $\dfrac{3^{-2}a}{5a^{-3}}$

10. Divide: $\dfrac{2x^3 + 15x^2 - 4}{2x - 1}$

11. Multiply: $(2x - 9)(x + 5)$

12. Subtract: $\begin{array}{r} 12x^2 + 8x - 1 \\ 4x^2 + 9x + 8 \end{array}$

13. Add: $(2x^2 - 3x - 7) + (7x^2 - 12x - 21)$

14. Is "$\frac{1}{2}$" the solution to the equation $8x - 7 = 6x - 5$?

15. Find H: $H = 7xy + 3y$ when $x = 4$, $y = -2$

16. Name the underlined: 1<u>2</u>xy _____

    _____

Chapter 10
Addition & Subtraction of Rational Expressions

Addition of Fractions with the Same Denominator - if the denominators of the rational expressions are the same, add or subtract the numerators and keep the same denominator.

Examples

a. $\dfrac{4}{x+3} + \dfrac{5}{x+3} = \dfrac{9}{x+3}$ b. $\dfrac{7}{x-1} - \dfrac{2}{x-1} = \dfrac{5}{x-1}$ c. $\dfrac{6}{x+5} - \dfrac{-2}{x+5} = \dfrac{6-(-2)}{x+5} = \dfrac{8}{x+5}$

Add or subtract

1. $\dfrac{6}{x+2} + \dfrac{5}{x+2}$

2. $\dfrac{9}{x-6} + \dfrac{3}{x-6}$

3. $\dfrac{7}{2x+1} + \dfrac{4}{2x+1}$

4. $\dfrac{2b}{x-5} + \dfrac{3b}{x-5}$

5. $\dfrac{3y}{x+10} + \dfrac{7y}{x+10}$

6. $\dfrac{4x}{2x+5} + \dfrac{3x}{2x+5}$

7. $\dfrac{9a}{x-3} + \dfrac{2a}{x-3}$

8. $\dfrac{6}{x+7} + \dfrac{-4}{x+7}$

9. $\dfrac{3}{x-4} + \dfrac{-2}{x-4}$

10. $\dfrac{8}{3y+5} - \dfrac{2}{3y+5}$

11. $\dfrac{7}{x-4} - \dfrac{1}{x-4}$

12. $\dfrac{8}{x+3} - \dfrac{2}{x+3}$

13. $\dfrac{6}{2x+1} - \dfrac{-2}{2x+1}$

14. $\dfrac{10a}{x-6} - \dfrac{-7a}{x-6}$

15. $\dfrac{9}{c} + \dfrac{2}{c}$

16. $\dfrac{3}{y} - \dfrac{7}{y}$

17. $\dfrac{11}{2b-3} + \dfrac{-4}{2b-3}$

18. $\dfrac{6}{7b} - \dfrac{-5}{7b}$

19. $\dfrac{-5}{x+3} - \dfrac{-3}{x+3}$

20. $\dfrac{-6}{4x+1} + \dfrac{-5}{4x+1}$

21. $\dfrac{3y}{z} - \dfrac{4y}{z}$

22. $\dfrac{9x}{x-7} - \dfrac{2x}{x-7}$

23. $\dfrac{8x}{3b} - \dfrac{7x}{3b}$

24. $\dfrac{5}{2a-7} + \dfrac{-2}{2a-7}$

More Addition of Fractions with the Same Denominator - if one or both of the numerators are binomials, combine any like terms of the numerators. Note: in subtraction problems, all of the terms following the negative sign must be changed.

Examples

a. $\dfrac{6x+1}{x+7} + \dfrac{4x-5}{x+7} = \dfrac{6x+1+4x-5}{x+7} = \dfrac{10x-4}{x+7}$ b. $\dfrac{3x+9}{2x-1} - \dfrac{2x-8}{2x-1} = \dfrac{3x+9-(2x-8)}{2x-1} = \dfrac{x+17}{2x-1}$

Add or subtract

1. $\dfrac{3a+2}{a+3} + \dfrac{2a+6}{a+3}$

2. $\dfrac{x+3}{x-4} + \dfrac{2x+5}{x-4}$

3. $\dfrac{3x+1}{x+9} + \dfrac{5x-3}{x+9}$

4. $\dfrac{4x-2}{4x+9} + \dfrac{2x+3}{4x+9}$

5. $\dfrac{2x+7}{x+2} - \dfrac{x+3}{x+2}$

6. $\dfrac{6x+5}{x-8} - \dfrac{x-4}{x-8}$

7. $\dfrac{4x-3}{x-10} - \dfrac{2x+3}{x-10}$

8. $\dfrac{x+6}{5x} + \dfrac{x-3}{5x}$

9. $\dfrac{2x-7}{6b} - \dfrac{3x-4}{6b}$

10. $\dfrac{4x-7}{4x+3} - \dfrac{3x-4}{4x+3}$

11. $\dfrac{6x+1}{6a} - \dfrac{6x-1}{6a}$

12. $\dfrac{x-7}{x+1} + \dfrac{x+2}{x+1}$

13. $\dfrac{4x-3}{7} - \dfrac{3x-3}{7}$

14. $\dfrac{x-6}{x-5} + \dfrac{x-3}{x-5}$

15. $\dfrac{x-7}{3x-5} - \dfrac{x+2}{3x-5}$

16. $\dfrac{8z+1}{7} + \dfrac{3z-1}{7}$

17. $\dfrac{6x-7}{5a} + \dfrac{-3x}{5a}$

18. $\dfrac{6b-5}{9} - \dfrac{5b-5}{9}$

Addition of Fractions with Similar Denominators - if the denominators have the same terms but opposite signs, **a)** put the denominators in the same order, **b)** factor out a negative one (-1), then, **c)** move the negative one (-1) to the numerator and simplify.

Examples

a. $\dfrac{4}{x-5} + \dfrac{3}{-x+5}$

$= \dfrac{4}{x-5} + \dfrac{3}{-1(x-5)}$ factor out a (-1)

$= \dfrac{4}{x-5} + \dfrac{-1(3)}{x-5} = \dfrac{1}{x-5}$ simplify

b. $\dfrac{x+9}{3x-1} + \dfrac{4x-3}{1-3x}$

$= \dfrac{x+9}{3x-1} + \dfrac{4x-3}{-1(3x-1)}$ factor out a (-1)

$= \dfrac{x+9}{3x-1} + \dfrac{-1(4x-3)}{3x-1}$ move (-1) up

$= \dfrac{x+9-4x+3}{3x-1} = \dfrac{-3x+12}{3x-1}$ simplify

Factor out a negative one then simplify

1. $\dfrac{5}{x-3} + \dfrac{6}{3-x}$

2. $\dfrac{7}{x-5} + \dfrac{2}{5-x}$

3. $\dfrac{9}{2x-3} + \dfrac{2}{3-2x}$

4. $\dfrac{4a+2}{x-5} + \dfrac{3a-4}{5-x}$

5. $\dfrac{3x+5}{3x-2} + \dfrac{2x+5}{2-3x}$

6. $\dfrac{8x-3}{6-x} + \dfrac{4x+1}{x-6}$

7. $\dfrac{6x+5}{2-x} + \dfrac{6x+4}{x-2}$

8. $\dfrac{3a-1}{10-6x} + \dfrac{4a-2}{6x-10}$

9. $\dfrac{x+7}{x-3} + \dfrac{x+6}{3-x}$

10. $\dfrac{4}{2x-1} + \dfrac{7}{1-2x}$

11. $\dfrac{6x}{x-9} + \dfrac{2x}{9-x}$

12. $\dfrac{7}{3x-7} + \dfrac{2x+3}{7-3x}$

Subtraction of Fractions with Similar Denominators - if the denominators have the same terms but opposite signs, **a)** put the denominators in the same order, **b)** factor out a negative one -1), **c)** combine the (-1) with the negative sign between the fractions to make the problem addition, then, **d)** combine like terms.

Example

$$\frac{3x-2}{x-4} - \frac{x+5}{4-x} = \frac{3x-2}{x-4} - \frac{x+5}{-1(x-4)} = \frac{3x-2}{x-4} + \frac{x+5}{x-4} = \frac{4x+3}{x-4}$$

Factor out a negative one then simplify

1. $\dfrac{4x+2}{x-6} - \dfrac{x+6}{6-x}$

2. $\dfrac{3x+1}{2x-1} - \dfrac{2x-3}{1-2x}$

3. $\dfrac{5x+4}{3x-2} - \dfrac{x+5}{2-3x}$

4. $\dfrac{5}{x-3} - \dfrac{2}{3-x}$

5. $\dfrac{7}{x-2} - \dfrac{4}{2-x}$

6. $\dfrac{3x}{x-9} - \dfrac{3x}{9-x}$

7. $\dfrac{5x-7}{2x-y} - \dfrac{2x-7}{y-2x}$

8. $\dfrac{6x+1}{3y-b} - \dfrac{3x}{b-3y}$

9. $\dfrac{4x-4}{x-5} - \dfrac{-10}{5-x}$

Finding the Least Common Denominator (LCD) -if the denominators in an addition or subtraction problem are not the same, then a least common denominator must be determined. **The least common denominator is the smallest term that all the denominators of the problem will divide into evenly.** Some problems have LCD's that are easier to determine than others and finding their LCD does not require a method.

Examples

a. $\dfrac{x}{5} + \dfrac{2a}{4}$ b. $\dfrac{9}{4x} + \dfrac{2}{x}$ c. $\dfrac{6}{ab} - \dfrac{4}{bc}$ d. $\dfrac{3y}{4} - \dfrac{5y}{6}$

The LCD is 20   The LCD is 4x   The LCD is abc   The LCD is 12

Find the LCD

1. $\dfrac{a}{3} + \dfrac{3a}{4}$

2. $\dfrac{4x}{7} - \dfrac{2x}{6}$

3. $\dfrac{2x-1}{2} + \dfrac{3x}{8}$

4. $\dfrac{4x-1}{4} + \dfrac{x+1}{12}$

5. $\dfrac{6x}{5} - \dfrac{x-3}{4}$

6. $\dfrac{5a}{16} + \dfrac{3a-2}{2}$

7. $\dfrac{6}{7a} + \dfrac{a+3}{a}$

8. $\dfrac{5x}{b} - \dfrac{3x+2}{4b}$

9. $\dfrac{2b}{3x} + \dfrac{3b}{x}$

10. $\dfrac{4x}{y} + \dfrac{2x+3}{3y}$

11. $\dfrac{3b+2}{xy} - \dfrac{4b}{yz}$

12. $\dfrac{5x-4}{ab} - \dfrac{-15}{bc}$

13. $\dfrac{16}{ax} - \dfrac{x+6}{cx}$

14. $\dfrac{4b}{ra} + \dfrac{3b}{rx}$

15. $\dfrac{x+3}{6} + \dfrac{x-2}{4}$

16. $\dfrac{5x+2}{8} - \dfrac{14}{3}$

17. $\dfrac{x}{7} + \dfrac{2x}{5}$

18. $\dfrac{7b}{6x} - \dfrac{3}{x}$

19. $\dfrac{x-6}{tu} - \dfrac{-5}{rt}$

20. $\dfrac{13}{10} + \dfrac{x+6}{2}$

21. $\dfrac{2b-1}{ac} - \dfrac{5b}{cd}$

22. $\dfrac{-3x}{5y} + \dfrac{-7x}{15y}$

23. $\dfrac{3c}{4x} + \dfrac{2x}{x}$

24. $\dfrac{5x}{2} + \dfrac{7x}{3}$

Tht Factoring Method Used to Find the LCD - to find the LCD when it is not obvious, follow these steps: **1)** Factor each denominator to its <u>prime factors</u>. **2)** Use each factor the <u>greatest number of times it appears in any one denominator</u>.

Examples

a. $\dfrac{5}{3a} - \dfrac{2}{a^2}$

(3)(a)   (a)(a)

3   (a)(a)

The LCD is $3a^2$

b. $\dfrac{3}{4x^2} + \dfrac{5b}{2xy}$

(2)(2)(x)(x)   (2)(x)(y)

(2)(2)(x)(x)   (y)

The LCD is $4x^2y$

c. $\dfrac{3}{2x^2} + \dfrac{4}{3x} - \dfrac{2b}{x^3}$

(2)(x)(x)   (3)(x)   (x)(x)(x)

(2)   (3)   (x)(x)(x)

The LCD is $6x^3$

Find the LCD

1. $\dfrac{16}{x^2} + \dfrac{3}{5x}$

2. $\dfrac{3}{10} - \dfrac{-5}{2x^2}$

3. $\dfrac{x+1}{4x^2} - \dfrac{x+3}{12xy}$

4. $\dfrac{3x-2}{x^3} - \dfrac{x+2}{3x}$

5. $\dfrac{3x}{7x^3} - \dfrac{x+2}{14xy}$

6. $\dfrac{3x+5}{3ab} + \dfrac{7x}{4a^2}$

7. $\dfrac{-11}{5x^2} - \dfrac{2}{3x}$

8. $\dfrac{7y}{6x} + \dfrac{4}{x^2}$

9. $\dfrac{2x}{5ab} + \dfrac{5x}{4b^2}$

10. $\dfrac{9x}{2y} + \dfrac{x+2}{3y^3} - \dfrac{x-1}{y^2}$

11. $\dfrac{5}{2a^2} - \dfrac{-7}{ab} + \dfrac{11}{8b^3}$

12. $\dfrac{4}{2xy} - \dfrac{x-3}{6x^2y} + \dfrac{x+1}{3xy^2}$

13. $\dfrac{5a}{2y^3} + \dfrac{23}{6y} + \dfrac{4a}{5}$

14. $\dfrac{7x}{3a^2} - \dfrac{5x}{4a^3} - \dfrac{2x}{5}$

15. $\dfrac{x-3}{2xy^2} + \dfrac{5x}{6y^2} + \dfrac{x+1}{3x^2}$

More Find the LCD Problems - two types of problems are shown in the examples.
1) In some problems, the denominators are factored, and the factors that appear most often in any one denominator make up the LCD.
2) In other problems the denominators are monomials or binomials that cannot be factored, and the LCD usually will be the product of both denominators.

Examples

1a) $\dfrac{x+1}{x^2-9} - \dfrac{5}{4x-12}$

$(x-3)(x+3)\quad 4(x-3)$

$(x-3)(x+3)\quad\quad 4$

The LCD is
$4(x-3)(x+3)$

1b) $\dfrac{6}{x^2-6x+5} + \dfrac{4}{(x-1)^2}$

$(x-5)(x-1)\quad (x-1)(x-1)$

$(x-5)\quad\quad (x-1)(x-1)$

The LCD is
$(x-5)(x-1)(x-1)$

2a) $\dfrac{3}{x+2} - \dfrac{5}{x}$

The LCD is
$x(x+2)$

2b) $\dfrac{6}{x-3} + \dfrac{2}{x+7}$

The LCD is
$(x-3)(x+7)$

Find the LCD

1. $\dfrac{6}{(x+2)^2} + \dfrac{3}{3x+6}$

2. $\dfrac{1}{x^2-12x+36} - \dfrac{4}{3x-18}$

3. $\dfrac{x}{4x-20} + \dfrac{4}{x^2-25}$

4. $\dfrac{3}{x-4} - \dfrac{2}{5}$

5. $\dfrac{4x}{x+3} - \dfrac{x+3}{x+6}$

6. $\dfrac{7}{5x} + \dfrac{x-1}{x+2}$

7. $\dfrac{6}{x^2-16} + \dfrac{x}{3x+12}$

8. $\dfrac{z-6}{z+5} - \dfrac{2}{3z}$

9. $\dfrac{3x}{x^2+4x+4} - \dfrac{6}{2x^2+4x}$

10. $\dfrac{x-4}{x+1} - \dfrac{7}{x+3}$

11. $\dfrac{7}{x+6} + \dfrac{3x}{2}$

12. $\dfrac{7}{(x+5)^2} + \dfrac{x}{3x^2+15x}$

Addition and Subtraction of Rational Expressions - take these steps: **1)** Determine the LCD. **2)** Determine what each denominator must be multiplied by to get the LCD. **3)** Multiply the numerator and denominator of each fraction by the polynomial found in step 2. **4)** Multiply and simplify to the lowest terms.

Examples

a. $\frac{x}{2} - \frac{3}{5}$ LCD: 10

$\left(\frac{5}{5}\right)\frac{x}{2} - \left(\frac{2}{2}\right)\frac{3}{5}$

$\frac{5x}{10} - \frac{6}{10} = \frac{5x-6}{10}$

b. $\frac{3}{x-y} - \frac{2}{x}$ LCD: x(x-y)

$\frac{x}{x} \bullet \frac{3}{(x-y)} - \left(\frac{x-y}{x-y}\right)\frac{2}{x}$

$\frac{3x}{x(x-y)} - \frac{2x-2y}{x(x-y)}$

$\frac{3x-2x+2y}{x(x-y)} = \frac{x+2y}{x(x-y)}$

c. $\frac{3}{x+2} + \frac{4}{x+4}$ LCD:(x+2)(x+4)

$\left(\frac{x+4}{x+4}\right)\frac{2}{x+2} + \left(\frac{x+2}{x+2}\right)\frac{4}{x+4}$

$\frac{2x+8}{(x+4)(x+2)} + \frac{4x+8}{(x+2)(x+4)}$

$\frac{6x+16}{(x+4)(x+2)}$

Add or Subtract

1. $\frac{3}{4} - \frac{4x}{5}$

2. $\frac{3}{x} + \frac{4}{3x}$

3. $\frac{5}{a+b} + \frac{4}{a}$

4. $\frac{5}{3x} + \frac{4}{2x+3}$

5. $\frac{6}{x+7} + \frac{2}{x+1}$

6. $\frac{4}{a-4} - \frac{5}{a-3}$

7. $\dfrac{3}{x-1} - \dfrac{5}{x}$

8. $\dfrac{7}{4x} - \dfrac{3}{x}$

9. $\dfrac{3}{x-4} + \dfrac{2}{5}$

10. $\dfrac{x}{4x-3} + \dfrac{5}{x+2}$

11. $\dfrac{4y}{5a} + \dfrac{3}{a}$

12. $\dfrac{5}{x-2} - \dfrac{4}{x-5}$

13. $\dfrac{3b}{x} + \dfrac{b}{x-3}$

14. $\dfrac{2}{3x-4} + \dfrac{3}{2x+1}$

15. $\dfrac{7}{4y} - \dfrac{2x}{yz}$

16. $\dfrac{6x}{x+10} - \dfrac{2x}{3}$

More Addition and Subtraction of Rational Expressions - these two types of problems have either three terms or require multiplication using the FOIL method.

There is another method for adding and subtracting rational expressions which uses shortcuts. Take these steps: 1) Divide each denominator into the LCD. **2)** Multiply each numerator by its denominator's quotient, (found in step 1). **3)** Multiply and simplify.

Examples

a. $\dfrac{3}{x^2} + \dfrac{5}{x^3} - \dfrac{2}{3x}$ LCD: $3x^2$

$\dfrac{3x(3)+3(5)-x^2(2)}{3x^3}$ ← step 1 →

$\dfrac{9x+15-2x^2}{3x^3}$

b. $\dfrac{x-5}{x+1} + \dfrac{x+2}{x+3}$ LCD: $(x+1)(x+3)$

$\dfrac{(x+3)(x-5)+(x+1)(x+2)}{(x+1)(x+3)}$

$\dfrac{(x^2-5x+3x-15)+(x^2+2x+1x+2)}{(x+1)(x+3)}$

$\dfrac{2x^2+5x-13}{(x+1)(x+3)}$

Add or Subtract (use the first method or try the shortcut method described above)

1. $\dfrac{3}{x^2} + \dfrac{4}{3x} - \dfrac{2}{x^3}$

3. $\dfrac{x+5}{x+2} + \dfrac{x-8}{x-6}$

2. $\dfrac{x+4}{x-9} + \dfrac{x-3}{x+1}$

4. $\dfrac{2x-1}{x+1} + \dfrac{x-4}{3x-1}$

5. $\dfrac{4}{b^2} + \dfrac{2}{b^3} - \dfrac{5}{7b}$

6. $\dfrac{2}{3a} + \dfrac{5}{4a^2} + \dfrac{1}{6a^3}$

7. $\dfrac{x+6}{x+3} + \dfrac{x+1}{x+4}$

8. $\dfrac{x-8}{x-4} + \dfrac{x+2}{x-6}$

9. $\dfrac{3}{10y^3} - \dfrac{5}{4y} - \dfrac{8}{5y^2}$

10. $\dfrac{5}{2a} - \dfrac{9}{4a^2} + \dfrac{1}{6a^3}$

11. $\dfrac{x-7}{2x-3} + \dfrac{x+1}{4x-1}$

12. $\dfrac{1}{2x^2} + \dfrac{6}{7x^3} - \dfrac{3}{4x}$

More Addition and Subtraction of Rational Expressions - these two types of problems are: **1)** subtraction problems where binomials are multiplied by binomials. The binomials must be multiplied first, then change the signs inside the parentheses following the negative (subtraction) sign. **2)** problems where the denominators need to be factored before determining the LCD.

Examples

1) $\dfrac{x-1}{x-2} - \dfrac{x+3}{x+6}$ LCD: $(x-2)(x+6)$

$\left(\dfrac{x+6}{x+6}\right)\dfrac{x-1}{x-2} - \left(\dfrac{x-2}{x-2}\right)\dfrac{x+3}{x+6}$

$\dfrac{x^2-1x+6x-6}{(x+6)(x-2)} - \dfrac{x^2+3x-2x-6}{(x-2)(x+6)}$

$\dfrac{x^2+5x-6}{(x+6)(x-2)} + \dfrac{-(x^2+1x-6)}{(x+6)(x-2)}$

$\dfrac{x^2+5x-6-x^2-1x+6}{(x+6)(x-2)} = \dfrac{4x}{(x+6)(x-2)}$

2) $\dfrac{7}{x^2+6x+9} - \dfrac{5}{2x+6}$

$\dfrac{7}{(x+3)(x+3)} - \dfrac{5}{2(x+3)}$ LCD: $2(x+3)(x+3)$

$\dfrac{2(7) \quad -(x+3)5}{2(x+3)(x+3)}$

$\dfrac{14-5x-15}{2(x+3)(x+3)}$

$\dfrac{-1-5x}{2(x+3)(x+3)}$

Add or Subtract (both methods are illustrated in the examples)

1. $\dfrac{x-1}{x+4} - \dfrac{x+2}{x-6}$

2. $\dfrac{5}{x^2+14x+49} - \dfrac{2}{3x+21}$

3. $\dfrac{x+3}{x-3} - \dfrac{x+5}{x-5}$

4. $\dfrac{5}{2x+10} + \dfrac{3}{x^2-25}$

5. $\dfrac{3x+2}{x-2} - \dfrac{x+5}{x-3}$

6. $\dfrac{x+5}{x+2} - \dfrac{x-6}{x-8}$

7. $\dfrac{2}{(x+5)^2} - \dfrac{1}{x^2+7x+10}$

8. $\dfrac{4}{x^2+2x-15} + \dfrac{7}{x^2-2x-3}$

Rational Equations - to solve rational equations, follow steps similar to those used in adding and subtracting rational expressions. **Note:** in the examples, after the LCD has been used, it is no longer needed and is dropped from the problem. Again, in the examples both methods will be illustrated.

Examples

a. $\dfrac{x}{6} = \dfrac{3}{5}$ LCD: 30

$$\left(\dfrac{5}{5}\right)\dfrac{x}{6} = \left(\dfrac{6}{6}\right)\dfrac{3}{5}$$

$$\dfrac{5x}{30} = \dfrac{18}{30} \quad \text{(drop the LCD)}$$

$$5x = 18$$

$$x = \dfrac{18}{5}$$

b. $\dfrac{3}{5} = \dfrac{x+1}{x-3}$ LCD $5(x-3)$

$$\dfrac{3(x-3)}{5(x-3)} = \dfrac{5(x+1)}{} \quad \text{(drop the LCD)}$$

$$3x - 9 = 5x + 5$$
$$-3x \quad -5 \quad -3x \quad -5$$
$$\dfrac{-14}{2} = \dfrac{2x}{2}$$
$$-7 = x$$

Solve the Rational Equations

1. $\dfrac{4}{3} = \dfrac{x}{7}$

2. $\dfrac{5}{6} = \dfrac{x+2}{x+1}$

3. $\dfrac{2}{3} = \dfrac{x+1}{x+3}$

4. $\dfrac{3}{5} = \dfrac{x-4}{x-5}$

5. $\dfrac{x}{3} = \dfrac{2}{7}$

6. $\dfrac{x+4}{x-7} = \dfrac{2}{5}$

7. $\dfrac{x-6}{x+4} = \dfrac{3}{2}$

8. $\dfrac{-7}{8} = \dfrac{4}{x}$

9. $\dfrac{-3}{4} = \dfrac{x}{5}$

10. $\dfrac{7}{2} = \dfrac{x-8}{x-4}$

11. $\dfrac{1}{-6} = \dfrac{-2}{x}$

12. $\dfrac{x-3}{x+9} = \dfrac{4}{7}$

More Solving Rational Equations - these equations have denominators that are binomials. Use the same steps that were used on the previous page. Again, two different methods are used.

Examples

a. $\dfrac{3}{x+2} = \dfrac{2}{x+4}$ LCD: $(x+2)(x+4)$

$\left(\dfrac{x+4}{x+4}\right)\dfrac{3}{x+2} = \left(\dfrac{x+2}{x+2}\right)\dfrac{2}{x+4}$

$\dfrac{3x+12}{(x+4)(x+2)} = \dfrac{2x+4}{(x+2)(x+4)}$ drop LCD

$3x + 12 = 2x + 4$
$-2x - 12 \quad -2x - 12$
$\quad x \quad = \quad -8$

b. $\dfrac{x+2}{x+4} = \dfrac{x-6}{x+1}$ LCD: $(x+4)(x+1)$

$\dfrac{(x+1)(x+2)}{(x+4)(x+1)} = \dfrac{(x+4)(x-6)}{}$ drop LCD

$x^2 + 2x + 1x + 2 = x^2 - 6x + 4x - 24$

$x^2 + 3x + 2 = x^2 + 4x - 24$
$\underline{-x^2\ -3x\ +24} \quad \underline{-x^2\ -3x\ +24}$
$\quad +26 = \quad x$

Solve the Rational Equations

1. $\dfrac{4}{x+3} = \dfrac{5}{x-1}$

2. $\dfrac{6}{x+7} = \dfrac{2}{x+2}$

3. $\dfrac{x+3}{x+5} = \dfrac{x+2}{x-1}$

4. $\dfrac{x-3}{x-7} = \dfrac{x+2}{x+1}$

5. $\dfrac{x+2}{x+3} = \dfrac{x+1}{x+4}$

6. $\dfrac{3}{x-3} = \dfrac{-1}{x+8}$

7. $\dfrac{x-3}{x-4} = \dfrac{x-5}{x+2}$

8. $\dfrac{x+5}{x+2} = \dfrac{x-7}{x-8}$

9. $\dfrac{4}{x-1} = \dfrac{6}{x+2}$

10. $\dfrac{7}{x-3} = \dfrac{6}{x-4}$

More Rational Equations - some of these rational equations have 3 terms, but the steps are no different from the steps taken in previous problems.

A term such as $\frac{x}{4}$ can be written as $\frac{1}{4}x$. Therefore, rational equations can take two different looks. Example: $\frac{2}{3}x + \frac{5}{2}x = \frac{5}{8}$ can be written as $\frac{2x}{3} + \frac{5x}{2} = \frac{5}{8}$.

Examples

a. $\frac{3}{x} + \frac{1}{2x} = \frac{1}{6}$ LCD: 6x

$\left(\frac{6}{6}\right)\frac{3}{x} + \left(\frac{3}{3}\right)\frac{1}{2x} = \frac{x}{x} \cdot \frac{1}{6}$

$\frac{18}{6x} + \frac{3}{6x} = \frac{1x}{6x}$ (drop LCD)

$18 + 3 = 1x$
$21 = x$

b. $\frac{1}{4}x + \frac{2}{5}x = 7$

$\frac{x}{4} + \frac{2x}{5} = \frac{7}{1}$ LCD: 20

$\frac{5(x) + 4(2x) = 20(7)}{20}$ (drop LCD)

$5x + 8x = 140$
$\frac{13x}{13} = \frac{140}{13}$
$x = \frac{140}{13}$

Solve the Rational Equations

1. $\frac{5}{x} + \frac{2}{3x} = \frac{1}{9}$

2. $\frac{6}{x} + \frac{2}{5x} = \frac{1}{10}$

3. $\frac{3}{5}x - \frac{1}{4}x = \frac{1}{2}$

4. $\frac{1}{4}x + \frac{2}{3}x = \frac{5}{8}$

Rational Equations Continued

5. $\dfrac{2x}{3} + \dfrac{3}{2} = \dfrac{5x}{6}$

8. $\dfrac{3}{2}x - \dfrac{1}{4}x = \dfrac{2}{3}$

6. $\dfrac{4}{3}x - \dfrac{1}{2}x = \dfrac{3}{4}$

9. $\dfrac{3x}{2} - \dfrac{3}{4} = \dfrac{5x}{8}$

7. $\dfrac{1}{5}x + \dfrac{2}{3}x = \dfrac{7}{10}$

10. $\dfrac{3}{2x} + \dfrac{4}{3x} = \dfrac{2}{3}$

More Rational Equations -these equations will become <u>quadratic equations</u>. To find the solutions (roots), factor the quadratic equation.

Examples

a. $\dfrac{x^2}{x-3} - \dfrac{12}{x-3} = 8$

$\dfrac{1(x^2)-1(12)=8(x-3)}{x-3}$ LCD: (x-3)

$\dfrac{x^2-12=8x-24}{x-3}$ ← drop LCD →

$x^2 - 8x - 12 + 24 = 0$
$x^2 - 8x + 12 = 0$
$(x-2)(x-6) = 0$
$x = 2$ or $x = 6$

b. $\dfrac{15}{x^2+x-20} = \dfrac{x}{x+5} - \dfrac{3}{x-4}$ LCD: (x+5)(x-4)

$\dfrac{15}{(x+5)(x-4)} = \left(\dfrac{x-4}{x-4}\right)\dfrac{x}{x+5} - \left(\dfrac{x+5}{x+5}\right)\dfrac{3}{x-4}$

$\dfrac{15}{(x+5)(x-4)} = \dfrac{x^2-4x}{(x+5)(x-4)} - \dfrac{3x+15}{(x+5)(x-4)}$

$15 = x^2 - 4x - (3x + 15)$
$15 = x^2 - 7x - 15$ (move the 15 over)
$0 = x^2 - 7x - 30$
$0 = (x-10)(x+3)$
$x = 10$ or $x = -3$

Solve the Rational Equations

1. $\dfrac{x^2}{x-2} - \dfrac{5}{x-2} = 4$

2. $\dfrac{x^2}{x-5} + \dfrac{-25}{x-5} = 3$

3. $\dfrac{9}{x^2-2x-35} = \dfrac{x}{x-7} + \dfrac{1}{x+5}$

4. $\dfrac{10}{x^2-4x-12} = \dfrac{x}{x+2} + \dfrac{5}{x-6}$

5. $\dfrac{19}{2x+1} + \dfrac{x^2}{2x+1} = 4$

8. $\dfrac{x}{x-3} = \dfrac{27}{x^2+3x-18} + \dfrac{7}{x+6}$

6. $\dfrac{-6}{x^2-8x-20} = \dfrac{3}{x+2} + \dfrac{x}{x-10}$

9. $\dfrac{4}{x+6} + \dfrac{x}{x+5} = \dfrac{-1}{x^2+11x+30}$

7. $\dfrac{x^2}{x+3} + \dfrac{11}{x+3} = 6$

10. $\dfrac{-8}{x^2+7x-8} + \dfrac{x}{x+8} = \dfrac{3}{x-1}$

Chapter 10 Review

Add, Subtract, or Solve

1. $\dfrac{3x-1}{2x+3} - \dfrac{3x+7}{2x+3}$

2. $\dfrac{3}{4} = \dfrac{x+1}{x-6}$

3. Find the LCD:
$\dfrac{2}{x^2-9} + \dfrac{5}{(x+3)^2}$

4. $\dfrac{3}{5} + \dfrac{7x}{2} = \dfrac{9}{10}$

5. $\dfrac{3}{x-8} + \dfrac{5}{8-x}$

6. $\dfrac{2}{x} = \dfrac{3}{5}$

7. $\dfrac{x}{x+3} + \dfrac{4}{x-3}$

8. $\dfrac{5}{2x} + \dfrac{4}{3x} = \dfrac{2}{3}$

9. $\dfrac{x-5}{x-4} = \dfrac{x+6}{x+8}$

10. $\dfrac{12x}{x+5} - \dfrac{7x}{x+5}$

11. $\dfrac{x}{x^2+7x+12} - \dfrac{2}{x+3}$

12. $\dfrac{3x}{5} - \dfrac{x}{3}$

13. $\dfrac{4}{a^3} + \dfrac{3}{a^2} - \dfrac{6}{a}$

14. $\dfrac{x^2}{3x-1} + \dfrac{45}{3x-1} = 5$

15. $\dfrac{7x}{x+9} - \dfrac{3x}{x+9}$

16. $\dfrac{1}{4}x + \dfrac{4}{5}x = \dfrac{1}{2}$

17. $\dfrac{2x-5}{3x} - \dfrac{2x+5}{3x}$

18. $\dfrac{12}{x+1} - \dfrac{-42}{x^2-5x-6} = \dfrac{x}{x-6}$

19. $\dfrac{3}{x-3} = \dfrac{-1}{x+8}$

20. $\dfrac{5x}{x-4} - \dfrac{2x}{4-x}$

Cumulative Review 8

Solve for the variable:

1. $3x(4x - 1) = 0$

2. $x^2 + 4x - 12 = 0$

3. Simplify: $\dfrac{(x-7)(2x+3)}{(2x+3)}$

4. Multiply: $\dfrac{2x^2+10x}{3} \times \dfrac{6}{x^2-25}$

5. Factor: $7ay - 21y - ax + 3x$

6. Factor: $10xy + 4y$

7. Solve and put on the number line:
   $7 - x \geq 3$

   ———|———
       0

8. Solve: $7x - 19 = 3x + 23$

9. Combine the like terms:
   $9x + 4x - x = 12 - 32 + 18$

10. Divide: $x - 1 \overline{\smash{\big)}\ x^2 + 7x - 8}$

11. Simplify: $\dfrac{x^{-5}y}{x^2 y^3 z^{-3}}$

12. Multiply: $(-5x^5yz^3)(-3xyz^4)$

13. Multiply: $-8a(a^2 - 6a + 5)$

Name the property

14. $10 \times 3 \times 2 = 3 \times 2 \times 10$ _____

15. $3 \bullet (5 \bullet 4) = (3 \bullet 5) \bullet 4$ _____

16. $1x = x$ _____

17. Simplify: $2^2 \times 5 + 6 \div 2 - 10 \div 2$

18. Simplify: $5 - (8 + 2\{9 - 4\} - 11) + 7$

Chapter 11
Solving Linear Equations

A linear equation can have <u>two variables</u>, and the exponents of each variable can only be a "1". $3x - 5y = -4$ is a linear equation. Two variables (x & y) and exponents of "1".

Forms of Linear Equations - there are two basic looks a linear equation can take, the <u>Standard Form</u> and the <u>Slope-Intercept Form</u>.

Standard Form
This form <u>always</u> begins with a positive number of x's, followed by a positive <u>or</u> negative number of y's, then an equal sign and an integer. ($3x - 5y = -4$) To put a linear equation in standard form, the terms of the equation must be moved to the proper position. (Any term that moves from one side of the equation to the other side of the equation must change signs.)

Change to Standard Form

a. $\quad 2y = 3x - 7$
$\quad\quad \underline{-3x \quad\quad -3x}$ Move 3x
$\quad -3x + 2y = -7$ Change -3x to 3x
$\quad\; 3x - 2y = +7$

Slope-Intercept Form
This form begins with a positive one y, (y), followed by an equal sign, next a positive or negative number of x's and then a rational number. Example: ($y = 4x - 7$) To put linear equations in slope-intercept form, the terms must be moved to the proper position. (If "y" has a numerical coefficient, the entire equation must by divided by y's numerical coefficient.)

Change to Slope-Intercept Form

b. $\quad 3x = 2y + 7$ {the 3x and 2y
$\quad \underline{-3x \;\; -2y \quad -3x - 2y}$ are moved}
$\quad \dfrac{-2y}{-2} = \dfrac{-3x}{-2} + \dfrac{+7}{-2}$ {divide by (-2)}
$\quad\quad y = \dfrac{3}{2}x - \dfrac{7}{2}$

Change to Standard Form

1. $4y + 3x = 1$

2. $2x = 5y - 6$

3. $y = -7x + 4$

Change to Slope-Intercept Form

4. $3y = 6x - 2$

5. $4x - 3 = 3y$

6. $5x - 8y = 2$

Changing Linear Equations to Standard Form & Slope-Intercept Form Continued -
when putting the equation into Standard Form, if there are fractions in the linear equation, multiply the entire equation by the least common denominator. (LCD)

Example

$y = \frac{3}{2}x - \frac{7}{2}$

$2(y = \frac{3}{2}x - \frac{7}{2})$ $2 \times y = 2y$; $\frac{2}{1} \times \frac{3x}{2} = \frac{6x}{2} = 3x$; $\frac{2}{1} \times \left(-\frac{7}{2}\right) = -\frac{14}{2} = -7$

$2y = 3x - 7$
$-3x + 2y = -7$
$3x - 2y = 7$

Change to Standard Form

7. $y = \frac{2}{5}x + \frac{1}{2}$

8. $4x = -5y + 2$

9. $3x = y + 3$

10. $\frac{1}{6}x = \frac{2}{3}y - 6$

Change to Slope-Intercept Form

12. $2x = y + 3$

13. $\frac{1}{4}x + y = \frac{2}{5}$ (don't eliminate the fractions)

14. $3y = 6x + 12$

15. $4x = 5y - 3$

Finding the Coordinates - the values of x and y that satisfy a linear equation are called <u>coordinates</u>. They are also called <u>ordered pairs</u> or <u>points</u>. In the equation, x + y = 6, if the value of x is 1, then the value of y has to be 5. These two numbers, (1,5), when paired together, are the coordinates.

To find a coordinate of a linear equation, substitute a number into the equation for x and solve the equation for y. Those values of x and y make up one coordinate (the x is always listed first and the y is listed second) Note: <u>any</u> values of "x" can be substituted into the equation. Unless otherwise directed, the student will determine those values.

Examples

Find and pair together two coordinates for the equation: 2x + y = 7 (substitute -3 and 0 for x)

$$2x + y = 7$$
$$2(-3) + y = 7$$
$$-6 + y = 7$$
$$+6 \quad\quad +6$$
$$y = 13$$

The coordinate is (-3,13)

$$2x + y = 7$$
$$2(0) + y = 7$$
$$0 + y = 7$$
$$y = 7$$

The coordinate is (0,7)

Find and pair together two coordinates for each equation, using the values for x in the parentheses

1. x + y = -3 (x = 2, x = 4)

2. x - y = 2 (x = -3, x = 0) Don't forget to change (-y) to (y).

3. 2x + y = -4 (x = 0, x = -2)

4. 3x - y = 3 (x = -3, x = 0)

Finding Coordinates Continued - if the equation is in <u>slope-intercept form</u>, you will still substitute the value of x into the formula to find the value of y.

Example

Find the coordinates for the equation: y = 3x - 4   (x = 3, x = 0)

$$y = 3x - 4$$
$$y = 3(3) - 4$$
$$y = 9 - 4$$
$$y = 5$$
Coordinate: (3,5)

$$y = 3x - 4$$
$$y = 3(0) - 4$$
$$y = 0 - 4$$
$$y = -4$$
Coordinate: (0,-4)

Find the coordinates, using the values in the parentheses

1. y = 2x + 3   (x = 1, x = -1)

2. 2x + y = 5   (x = 1, x = 3)

3. y = -3x + 5   (x = -3, x = 0)

4. y = -x - 6   (x = -5, x = -2)

5. x - y = 0   (x = 2, x = 5)

6. y = 2x + 4   (x = 0, x = 3)

Putting the Coordinates on the Coordinate Plane - the coordinate plane, often called a graph, is made up of two perpendicular lines, one horizontal, called the x-axis, and one vertical, called the y-axis.

On the x-axis, the values to the right of 0 are positive and the values to the left of 0 are negative. On the y-axis, the values above 0 are positive and the values below 0 are negative. The points where the two axis cross is called the origin, and is given the value of 0 on both axis.

To graph the coordinates, called plotting the points - **1)** Start at the origin, (0), and move on the x-axis the number of places indicated by the first number in the coordinate. **2)** From that place on the x-axis, move on the y-axis the number of places indicated by the second number in the coordinate. (when plotting "y", do not go back to the origin) **3)** At that place put a dot on the graph and label that dot.

Examples
Graph the coordinates, (-3,4) and (5,-2), on the coordinate plane

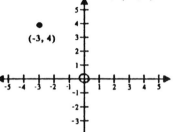

1) Move to (-3) on the x-axis.
2) Move to (4) on the y-axis.
3) Place a dot and label it (-3,4)

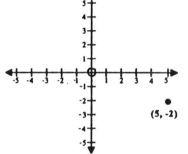

1) Move to (5) on the x-axis.
2) Move to (-2) on the y-axis.
3) Place a dot and label it (5,-2).

Graph the coordinates on the coordinate plane

1. (6,-2), (2, -1)

3. (3,4), (2,-5)

5. (3,-2), (-1,0), (2,4)

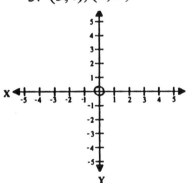

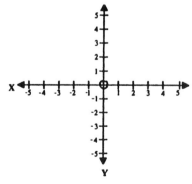

2. (-3,1), (-1,-4)

4. (-1,-2), (3,3)

6. (5-2), (0,2), (-1,-4)

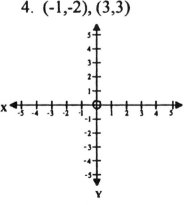

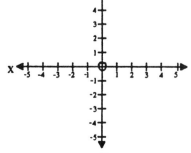

Graphing a Linear Equation - after finding three coordinates of a linear equation, and after plotting those coordinates (points) and connecting those points, you have now put the linear equation on the coordinate plane. This is called **graphing a linear equation.** The three points that have been plotted should represent a straight line.

Example
Graph the linear equation, $2x + y = 1$, using: $x = -2$, $x = 0$, $x = 3$

In this problem, it is easier to determine the values for "y" if you change $2x + y = 1$ to the slope-intercept form: $y = -2x + 1$

Find the coordinates
$y = -2x + 1$
$y = -2(-2) + 1 = 5$
$y = -2(0) + 1 = 1$
$y = -2(3) + 1 = -5$

Chart the coordinates

x	y
-2	5
0	1
3	-5

Graph the coordinates

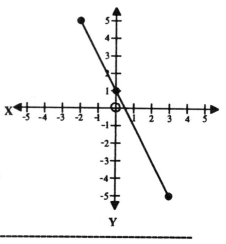

To work this problem, we have: 1) substituted the three values of "x" into the slope intercept form of the equation 2) put the coordinates on an <u>x and y chart</u> 3) put the coordinates on the graph 4) connected the 3 points.

Graph the linear equations

1. $x + y = 2$ (use $x = -3$, $x = 0$, $x = 3$) {use either standard form <u>or</u> slope-intercept form}

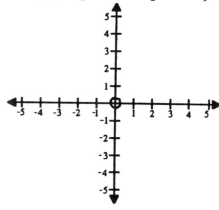

2. $2x - y = 3$ (use $x = 0$, $x = 2$, $x = 4$)

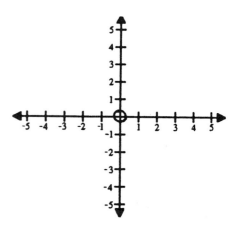

Graphing Linear Equations Continued - in graphing these equations, the student should determine what values to substitute into the equation for "x" to find "y". The student is also free to change the form when substituting. (standard or slope-intercept)

Graph the Linear equations

1. $3x - y = 4$

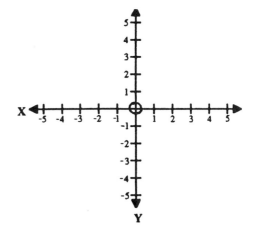

2. $y = 2x + 1$

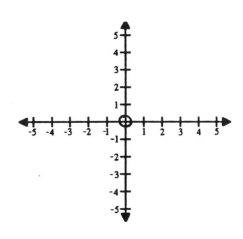

3. $x + y = -1$

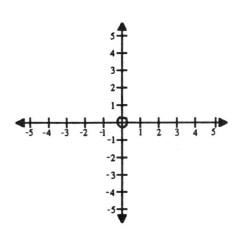

Solving Systems of Linear Equations by Graphing - the solution to two linear equations paired together (called <u>systems of linear equations</u> or <u>simultaneous equations</u>), is the point on the graph where the two equations cross. At that intersection, the same values of x and y will satisty <u>both</u> equations.

To solve simultaneous equations, 1) put both equations on the graph, extending each line far enought to make them cross, 2) estimate the value on the x-axis where the two lines cross, 3) estimate the value on the y-axis where the two lines cross. 4) These values of "x" and "y" make up the coordinate that is the solution to the simultaneous equations.

Example

Solve the <u>systems of linear equations</u> by graphing

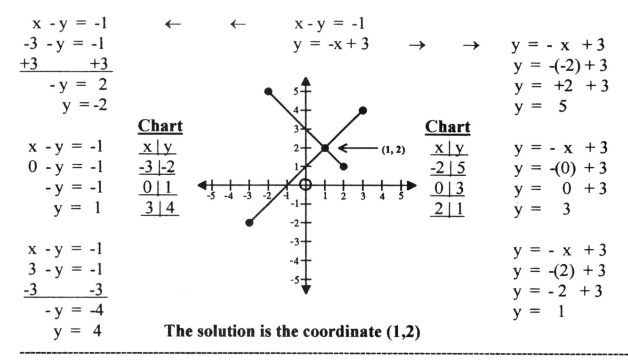

The solution is the coordinate (1,2)

Solve by graphing, choosing your own values for "x"

1. $x + y = 2$
   $y = x - 6$

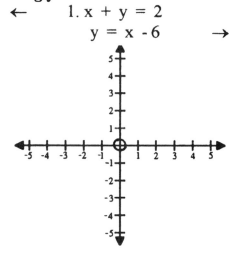

160

Solving Systems of Linear Equations by Graphing Continued

2. $x - y = 2$
   $y = -x - 4$

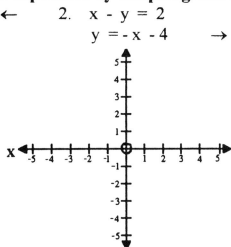

3. $2x + y = 3$
   $x - y = 6$

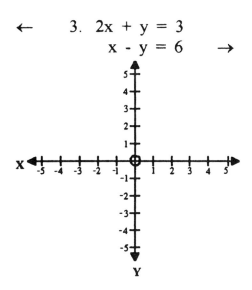

4. $x + y = 5$
   $y = 3x - 3$

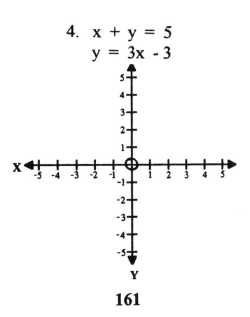

Solving Systems of Linear Equations by Graphing Continued

5. $3x - y = -2$
   $x + y = 2$

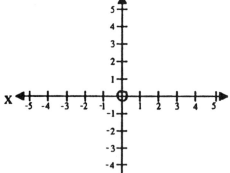

6. $2x + y = -4$
   $3x - y = -1$

7. $y = 2 - 3x$
   $y = 2x - 8$

Solving Systems of Linear Equations by Addition - this is the second method for solving simultaneous equations. This method can also be called <u>elimination</u>. In the addition method, the x's or the y's are <u>eliminated</u> by <u>addition</u> (or subtraction). Then the equation will only have one variable, and it can be solved.

Examples

a. $2x + y = 2$ (by addition)
    $\underline{3x - y = 8}$
    $5x = 10$
    $5 5$
    $x = 2$

b. $4x + y = -4$ (by subtraction)
    $\underline{\ominus 2x \oplus y = \ominus 2}$
    $2x = -6$
    $2 2$
    $x = -3$

Solve for x or y by addition or subtraction

1. $x + y = -8$
   $x - y = 4$

2. $3x - y = 2$
   $x + y = -6$

3. $2x + y = 5$
   $x + y = 4$

4. $3x - y = 5$
   $x - y = 1$

5. $3x - y = 1$
   $x + y = 7$

6. $x + 2y = 2$
   $-x + y = 7$

7. $x + 3y = 4$
   $x - 2y = -1$

8. $2x - y = -5$
   $x + y = -1$

9. $x - y = 2$
   $x + y = 6$

10. $3x + y = -5$
    $2x + y = 1$

11. $x + 5y = 2$
    $-x - 3y = 2$

12. $2x + 2y = -3$
    $x - 2y = -3$

More Solving Simultaneous Equations by Addition - after one of the two variables has been solved, that value is <u>substituted</u> into both of the equations to find the value of the other variable. This porcedure is called <u>Sub #1</u> and <u>Sub #2</u>.

Example
Solve for <u>both</u> variables by addition

First Step:	**Sub #1**	**Sub #2**
$3x - y = 5$	$3x - y = 5$	$2x + y = 15$
$\underline{2x + y = 15}$	$3(4) - y = 5$	$2(4) + y = 15$
$5x = 20$	$12 - y = 5$	$8 + y = 15$
$5 5$	$\underline{-12 -12}$	$\underline{-8 -8}$
$x = 4$	$-y = -7$	$y = 7$
	$y = 7$	

Since $y = 7$ in both Sub #1 and Sub #2, the coordinate (4, 7) is the solution

---

Solve the simultaneous equations by addition

1. $x + y = 4$
   $x - y = -8$

2. $2x - y = 4$
   $x + y = 5$

3. $-x + y = 1$
   $x + 2y = 2$

4. $x - 2y = 4$
   $x + 3y = -1$

5. $2x - y = 5$
   $x + y = 7$

6. $x - y = 6$
   $x + y = 2$

7. $x - 2y = -3$
   $3x + 2y = -1$

8. $x + 3y = 2$
   $-x + 2y = 3$

More Solving Simultaneous Equations by Addition - in some problems neither the x's nor the y's will be eliminated by addition or subtraction. When this happens, one of the equations can be multiplied by a number that will make the x's or y's disappear.

Example
Solve the simultaneous equation by addition
$$3x + 2y = 8$$
$$2x - y = 3$$

Multiply this equation by (2), which will cause the y's to be eliminated when the two equations are added.

$$2x - y = 3$$
$$2(2x - y = 3)$$
$$4x - 2y = 6$$

$$3x + 2y = 8$$
$$4x - 2y = 6$$
$$\frac{7x}{7} = \frac{14}{7}$$
$$x = 2$$

Sub #1
$$3x + 2y = 8$$
$$3(2) + 2y = 8$$
$$6 + 2y = 8$$
$$-6 \quad -6$$
$$\frac{2y}{2} = \frac{2}{2}$$
$$y = 1$$

Sub #2
$$2x - y = 3$$
$$2(2) - y = 3$$
$$4 - y = 3$$
$$-4 \quad -4$$
$$-y = -1$$
$$y = 1$$

Since y = 1 in both Sub #1 and Sub #2 the coordinate (2,1) is the solution

---

Solve the simultaneous equations by addition

1. $x - 2y = 4$
   $x + y = 1$

2. $x - y = 2$
   $x + 3y = 2$

3. $3x + y = 4$
   $2x - 3y = -23$

Solving Simultaneous Equations by Addition Continued - is some of these problems, for the x's or y's to be eliminated, the equation must be multiplied by <u>a negative number</u>.

4. $4x + y = -3$
    $x + 2y = 1$

5. $3x - 4y = 1$
    $x + y = -2$

6. $x + 5y = 6$
    $2x + y = 3$

7. $x - y = -1$
    $4x + 3y = 10$

More Solving Simultaneous Equations by Addition - sometimes both equations must be multiplied by a number to eliminate one of the variables. When this occurs:
1) multiply the first equation by the <u>second equation's "y" coefficient</u>. **2)** multiply the second equation by the <u>first equation's "y" coefficient</u>. **3)** If the <u>signs of "y" are the same</u>, one the the two equations must be multiplied by a <u>negative coefficient of "y"</u>. **4)** The equations can now be added to solve for "x".

Example
Solve the simultaneous equation by addition
$$3x + 2y = 7$$
$$4x + 5y = 7$$

Multiply the first equation by (5). → $5(3x + 2y = 7)$ → $15x + 10y = 35$
Multiply the second equation by (-2). → $-2(4x + 5y = 7)$ → $-8x - 10y = -14$
$$\frac{7x}{7} = \frac{21}{7}$$
$$x = 3$$

To complete the problem, find "y" by using the <u>Sub #1</u> and <u>Sub #2</u> procedure.

Solve the simultaneous equations by addition (include Sub #1, Sub #2 in your answer)

1. $3x + 2y = 4$
   $2x - 3y = 7$

2. $7x + 5y = 26$
   $2x + 3y = 9$

3. $3x + 5y = 27$
   $5x - 4y = 8$

Solving Simultaneous Equations by Addition Continued

4. $3x + 5y = 1$
    $4x + 3y = -6$

5. $3x + 7y = -28$
    $4x - 3y = 12$

6. $4x - 2y = 14$
    $3x + 5y = 30$

7. $6x + 3y = -12$
    $4x + 7y = -18$

Solving Simultaneous Equations by Substitution - This is the third method for solving simultaneous equations.

When one of the equations is in slope-intercept form, that value of "y" can be substituted into the other equation for "y". The equation can then be solved for "x". When "x" is found, find "y" by the Sub #1 and Sub #2 procedure.

Example

Solve the Simultaneous Equations by Substitution

$2x - y = 3$
$y = 4x + 3$ ← (This value of y, $4x + 3$ is substituted into the 1st equation.)

$2x - (4x + 3) = 3$
$2x - 4x - 3 = 3$
$-2x - 3 = 3$
$ +3 +3$
$\underline{-2x = 6}$
$-2 -2$
$x = -3$

Sub #1	Sub #2
$2x - y = 3$	$y = 4x + 3$
$2(-3) - y = 3$	$y = 4(-3) + 3$
$-6 - y = 3$	$y = -12 + 3$
$+6 +6$	$y = -9$
$-y = 9$	
$y = -9$	

Since y = -9 in both Sub #1 and Sub #2, the coordinate (-3, -9) is the solution

---

Solve by substitution (include Sub #1 and Sub #2 in your answer)

1. $3x + y = 2$
   $y = 2x - 8$

2. $x + y = -5$
   $y = x + 1$

169

Solving Simultaneous Equations by Substitution Continued

3. $4x + 3y = 2$
   $y = x + 3$

4. $x - y = 11$
   $y = 10 - 6x$

5. $x + 3y = 6$
   $y = x - 2$

More Solving Simultaneous Equations by Substitution - when <u>neither</u> of the equations is in <u>slope-intercept form,</u> before the substitution can take place, one of the equations must be changed from standard form to slope-intercept form.

Example

$x + 2y = 2$
$x - y = -7$ → $x - y = -7$
$\, -x \quad -x$
$-y = -x - 7$

$x + 2y = 2$ ← (substitute into) $y = x + 7$
$x + 2(x + 7) = 2$
$x + 2x + 14 = 2$
$3x + 14 = 2$
$ -14 \; -14$
$\dfrac{3x}{3} = \dfrac{-12}{3}$
$x = -4$

Sub #1
$x + 2y = 2$
$-4 + 2y = 2$
$+4 +4$
$\dfrac{2y}{2} = \dfrac{6}{2}$
$y = 3$

Sub #2
$x - y = -7$
$-4 - y = -7$
$+4 +4$
$-y = -3$
$y = 3$

Since $y = 3$ in both Sub #1 and Sub #2, the coordinate $(-4, 3)$ is the solution.

Solve the simultaneous equations by substitution. (include Sub #1 and Sub #2)

1. $x + 3y = 6$
$x - y = 2$

2. $x + y = -5$
$x - y = -1$

More Solving Simultaneous Equations By Substitution Continued

3. $3x + y = 2$
   $2x - y = 8$

4. $2x - 4y = 10$
   $3x + y = 1$

5. $2x - y = 4$
   $x + 3y = 2$

Solving Systems of Linear INEQUALITIES by Graphing - linear inequalities contain inequality symbols instead of equal signs. They are put on the coordinate plane in the same way that linear equations are, with this exception: If the greater than (>) or less than (<) symbols are used, the coordinates of that equation are connected with a dotted line.

The Solutions - The right or the left side of each line is shaded in to represent the points that will satisfy that equation. When the inequality is in standard form, the (>) and (≥) symbols indicate shading to the right of the line. When in standard form, the (<) and (≤) symbols indicate shading to the left of the line. The area shaded in by both equations contain the coordinates that are the solutions to the problem.

Example
Solve the systems of linear inequalities by graphing

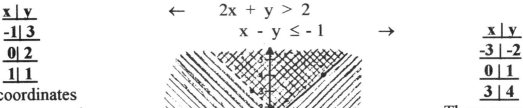

x	y
-1	3
0	2
1	1

These coordinates will be represented by a **dotted** line and the area to the **right** of the line will be shaded in.

x	y
-3	-2
0	1
3	4

These coordinates will be represented by a **solid** line and the area to the **left** of the line will be shaded in.

The solutions are all the possible coordinates shaded in twice. (the slice of pie on the top)

Solve by graphing and indicate the portion of the graph that contains the solutions

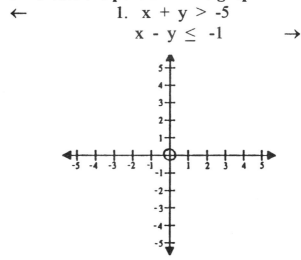

1. x + y > -5
    x - y ≤ -1

173

Solve the Simultaneous Linear Inequalities by Graphing Continued

2. $x + 3y \geq 6$
    $x - y > 2$

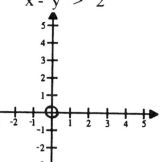

3. $3x + y < 2$
    $2x - y \geq 8$

4. $3x + 2y \leq -5$
    $2x + y \geq -4$

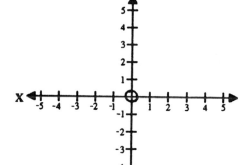

Chapter 11 Review

1. Solve for "x" and "y" by addition
   $3x + 2y = -3$
   $x - y = 4$

2. Solve by graphing and shade in the solution
   $4x + y < 2$
   $x - y \geq 4$

3. Solve for "x" and "y" by substitution
   $3x - y = 4$
   $y = 2x - 1$

4. Solve for "x" and "y" by addition
   $5x - 3y = 13$
   $3x + 2y = 4$

Chapter 11 Review Continued

5. Put in standard form
   a. $y = 4x + 6$
   b. $3y - 1 = -2x$

6. Put in slope-intercept form
   a. $3x + 2y = 7$
   b. $5 - 3y = 2x$

7. Solve for "x" and "y" by substitution
   $2x - y = 6$
   $3x + y = 4$

8. Solve by graphing
   $2x + y = 7$
   $2x - y = -3$

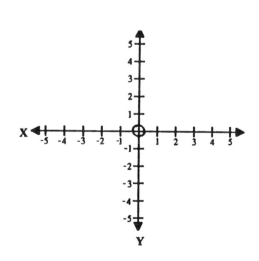

Cumulative Review 9

Add, Subtract, or Solve:

1. $\dfrac{x}{x+4} + \dfrac{2}{x-3}$

2. $\dfrac{4}{3x} + \dfrac{2}{x} = \dfrac{2}{3}$

3. $\dfrac{7a}{a-3} - \dfrac{2a}{a-3}$

4. $\dfrac{9}{x-6} + \dfrac{3}{6-x}$

5. $\dfrac{6x}{7} - \dfrac{x}{2}$

6. Solve: $(x-1)(2x+5) = 0$

7. Multiply: $\dfrac{5x^3z}{3xy^3} \bullet \dfrac{6y^2}{25xz}$

8. Divide: $\dfrac{x^2-3x-10}{8x} \div \dfrac{x^2-25}{16x}$

9. Factor: $2x^2 - 9x + 7$

10. $3x^3 - 12x$

11. $8ab - 3b$

12. Solve: $\dfrac{2x}{5} + 11 = 5$

13. Simplify: $-(ab)^0 =$

14. Multiply: $(4x^2y^4)^3$

15. Multiply: $4x^2 - x + 12$
    $\underline{3x - 2}$

Add or Subtract:

16. $7x^2 - x^2 - 9 + 5x - 7x - 1 + 4x^2 + 13 - x$

17. $(19x^2 - 13x + 5) - (15x^2 - 20x - 7)$

177

Chapter 12
Slope and Writing Equations

The slope of a line on the coordinate plane describes its <u>direction</u> and its <u>steepness</u>.
1. Every straight line has a <u>number that describes its slope</u>.
2. If the top of a line goes to the <u>right</u>, the slope is <u>positive</u>; if the top of the line goes to the <u>left</u>, the slope is <u>negative</u>.

Examples

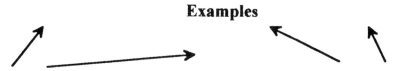

These lines have <u>positive</u> slopes   These lines have <u>negative</u> slopes

3. If the angle of the line is <u>steeper than 45 degrees</u>, the slope will be <u>greater than 1</u>.
4. If the line's steepness is <u>less than 45 degrees</u>, the slope will be <u>less that 1</u>. (a fraction)

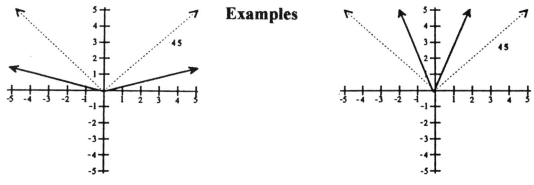

These slopes are <u>less than 1</u>   These slopes are <u>greater than 1</u>

Describe these slopes as <u>positive (+)</u> or <u>negative (-)</u>, <u>greater</u> than 1 or <u>less</u> than 1

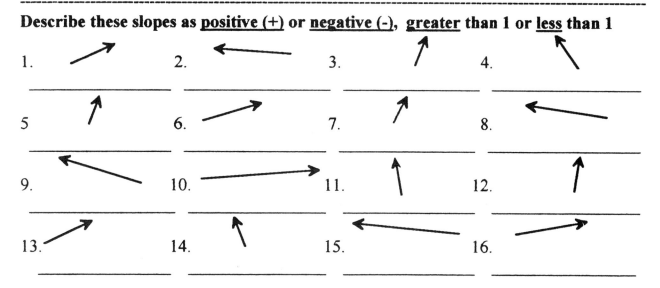

179

Finding the Slope of a Linear Equation (line) - there are 4 ways to find the slope of a linear equation. Which method to use will depend on what information is given.

Standard Form of a Linear Equation - when the equation is in this form ($3x + 4y = 2$), it is best to use the formula: $m = -\frac{A}{B}$. (The symbol for slope is the letter "m".) In this formula, "A" is the coefficient of "x" and "B" is the coefficient of "y". When these values are substituted into the formula, the slope becomes obvious.

Slope-Intercept Form of a Linear Equation - when the equation is in this form ($y = 2x - 7$), <u>The slope is the rational number in front of the variable, "x"</u>.

Examples

<u>Standard Form</u>		<u>Slope - Intercept Form</u>	
a. Find the slope	b. Find the slope	c. Find the slope	d. Find the slope
$3x + 4y = 2$	$5x - 2y = 6$	$y = -4x + 1$	$y = \frac{2}{3}x - 7$
$A = 3, B = 4$	$A = 5, B = -2$	The slope is (-4).	The slope is $\frac{2}{3}$
$m = -\frac{A}{B} = -\frac{3}{4}$	$m = -\frac{A}{B} = -\frac{5}{-2} = \frac{5}{2}$		

Find the slope (the method used should be determined by the <u>form</u> of the equation)

1. $2x + y = 7$
2. $6x - 5y = 2$
3. $4x + 3y = 5$
4. $y = 2x + 3$
5. $y = -3x - 4$
6. $y = \frac{1}{2}x - 6$
7. $5x - 2y = -3$
8. $8x - 6y = 4$
9. $y = 6x + 3$
10. $9x - 2y = -4$
11. $y = -\frac{5}{8}x - 6$
12. $x + 4y = 8$
13. $y = 5x + 9$
14. $y = \frac{5}{3}x + \frac{1}{2}$
15. $6x + 10y = 0$
16. $y = -7x - 3$
17. $5x - y = 2$
18. $x + y = 4$

Finding the Slope of a Linear Equation Continued

If two coordinates are given, the formula: $m = \frac{y_1 - y_2}{x_1 - x_2}$ **is used.** When using this formula, y_1 is the value of y in the first coordinate and y_2 is the value of y in the second coordinate. x_1 is the value of x in the first coordinate and x_2 is the value of x in the second coordinate. When these values are put into the formula and the numerators and denominators are simplified, the result will be the slope.

Example
Find the slope given two coordinates

(4,1), (2,-5)

$$m = \frac{y_1 - y_2}{x_1 - x_2} = \frac{1-(-5)}{4-2} = \frac{1+5}{2} = \frac{6}{2} = 3$$

Find the slope given the coordinates

1. (3,4), (-2,3)

2. (6,-2), (0,2)

3. (-2,5), (3,0)

4. (6,-1), (-1,-4)

5. (-6,-4), (-1,3)

6. (4,-6), (3,-2)

7. (-3,-4), (-1,6)

8. (-5,-1), (6,7)

9. (5,-1), (4,-3)

10. (-5,2), (1,6)

Finding the Slope of a Linear Equation Continued - if the linear equation is on the coordinate plane, the formula $m = \frac{rise}{run}$ can be used. Follow these steps: **1)** count the number of places from the <u>lowest coordinate</u> up to the level of the <u>highest coordinate</u>. Put that number in the formula across from "rise". **2)** From that place "run" sideways to the highest coordinate, counting the places. Put than number across from "run". **3)** Simplify the fraction if necessary and that will be the slope.

Examples
Find the slope of the lines on the coordinate planes, using the <u>rise over run</u> formula

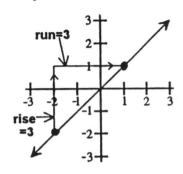
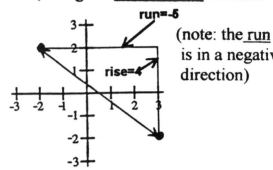

(note: the <u>run</u> is in a negative direction)

$$m = \frac{rise}{run} = \frac{3}{3} = 1 \qquad\qquad m = \frac{rise}{run} = \frac{4}{-5} = -\frac{4}{5}$$

Find the slope of the lines on the coordinate plane, using the <u>rise over run</u> formula

1. 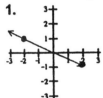   4. 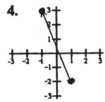   7. 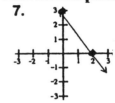   10. 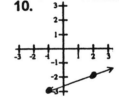   13.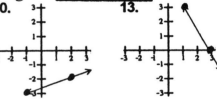

2. 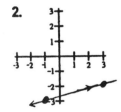   5. 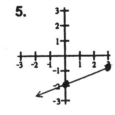   8. 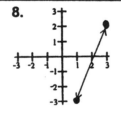   11. 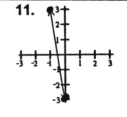   14.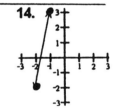

3. 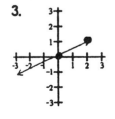   6. 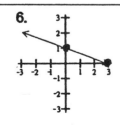   9. 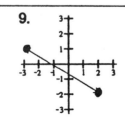   12. 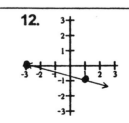   15.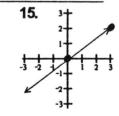

Writing Linear Equations - given certain information, a linear equation can be written. The slope-intercept formula is: **y = mx + b**. The 4 letters stand for the following: **y** is the coordinate "y", **m** is the slope, **x** is the coordinate "x", and **b** is the "y-intercept". To write an equation, you must have the slope and y-intercept, (**m** and **b**). They are put into the formula and a linear equation in slope-intercept form is written.

Examples
Write the equation in slope-intercept form

a. slope = -5
  y-intercept = 7
  y = mx + b
  y = -5x + 7

b. m = 8
  b = -3
  y = mx + b
  y = 8x - 3

Write the equation in slope intercept form, using the formula: y = mx + b

1. slope = -5, y-intercept = -2

2. slope = 3, y-intercept = 7

3. m = -6, b = 4

4. m = 10, b = -6

5. slope = -1, b = -6

6. m = 7, y-intercept = 2

7. slope = $\frac{4}{5}$, b = 1

8. m = 2, b = $-\frac{3}{2}$

9. slope = $-\frac{1}{5}$, y-intercept = $\frac{3}{4}$

10. slope = 9, b = -9

Writing an Equation Given the Slope and One Coordinate - even though the "y-intercept" is not given, one can still write a linear equation given the slope and one set of points.

Follow these steps: 1) substitute the values of "x" and "y" (from the coordinate given) into the slope-intercept formula for the "x" and "y", **2)** substitute the slope into the formula for "m", **3)** solve the formula for "b", the y-intercept, **4)** substitute the slope and the y-intercept into the formula and you have written a linear equation.

Example
Write an equation given the slope and one set of points

The problem: a. slope = 3, (4, -6)   b. (put the slope and the y-intercept
The formula: $y = mx + b$ into the formula again)
The substitution: $-6 = 3(4) + b$
Solve for **b**: $-6 = 12 + b$
                   $\underline{-12 \quad -12}$ $y = mx + b$
                   $-18 = \quad\quad b$ $y = 3x - 18$

Write the equation given the slope and one set of points

1. slope = 4, (3, -1)

2. slope = -7, (-2, 5)

3. m = 9, (-5, -3)

4. m = -1, (4, 8)

5. slope = 4, (-2, 3)

6. m = -3, (5, 0)

7. m = -1, (6, -3)

8. slope = 5, (-7, -2)

Writing an Equation Given Two Sets of Points - using the two sets of points, the <u>slope and the y-intercept</u> can be found.

Follow these steps: **1)** use the points to find the slope using $m = \frac{y_1 - y_2}{x_1 - x_2}$ **2)** put the slope and one set of points (it doesn't matter which one) into the slope-intercept formula to find the y-intercept **3)** put the slope and the y-intercept into the formula to write the equation

Example
Write the equation given two sets of points
(6,-7), (3,-1)

Find the slope	**Find the y-intercept**	**Write the equation**
$m = \frac{y_1 - y_2}{x_1 - x_2}$	y = mx + b	y = mx + b
	-7 = -2(6) + b	y = -2x + 5
$m = \frac{-7-(-1)}{6-3} = \frac{-7+1}{3} = \frac{-6}{3} = -2$	-7 = -12 + b	
	+12 +12	
	5 = b	

Write the equation given two sets of points

1. (4,5), (2,1)

2. (3, -2), (-3, -8)

3. (-1,7), (1,5)

Writing An Equation Given Two Sets of Points Continued

4. (-2,-1), (0,3)

5. (3,4), (5,-4)

6. (-3,6), (-5,2)

7. (6,10), (3,-5)

8. (3,-3), (7,5)

Writing Identical, Parallel, and Perpendicular Equations - given a linear equation, it is a simple procedure to write equations that are identical to, parallel to, and perpendicular to that linear equation. **To write a linear equation that is:**

Identical to another equation - multiply the entire equation by <u>any rational number</u>.

$$4x - 7y = 2 \rightarrow 2(4x - 7y = 2) \rightarrow 8x - 14y = 4$$

Parallel to another equation - do not change the coefficients (linear equations with the same coefficients will have the same slope; therefore, will be parallel) but <u>do change</u> the number to the right of the equal sign.

$$4x - 7y = 2 \rightarrow 4x - 7y = 3 \text{ (or any number \underline{but} 2)}$$

Perpendicular to another equation - <u>switch the "x" and "y" coefficients</u> and <u>change the sign</u> of the "y" coefficient.

$$4x - 7y = 2 \rightarrow 7x + 4y = 2 \text{ (or \underline{any real number})}$$

Write an equation identical, parallel, and perpendicular to the linear equations

1. $2x + 9y = 5$
Equal to:

Parallel to:

Perpendicular to:

2. $4x - 2y = -3$
Equal to:

Parallel to:

Perpendicular to:

3. $7x - 8y = 6$
Equal to:

Parallel to:

Perpendicular to:

4. $x - 6y = 3$
Equal to:

Parallel to:

Perpendicular to:

5. $10x - 9y = 8$
Equal to:

Parallel to:

Perpendicular to:

6. $6x + 3y = 4$
Equal to:

Parallel to:

Perpendicular to:

7. $4x + 5y = -8$
Equal to:

Parallel to:

Perpendicular to:

8. $7x - 12y = -6$
Equal to:

Parallel to:

Perpendicular to:

9. $2x + 5y = -7$
Equal to:

Parallel to:

Perpendicular to:

Chapter 12 Review

1. Find the slope using $m = \dfrac{y_1 - y_2}{x_1 - x_2}$ given the points $(3,-1), (6,-10)$

2. Write the equation given slope = 5, y-intercept = -9

3. Write the equations that are equal to:

   parallel to:

   and perpendicular to: $5x - 6y = 3$

4. Find the slope of $2x + 7y = 1$ using the formula: $m = -\dfrac{A}{B}$

5. Write the equation given the points $(-7,1), (-1,-5)$

6. Find the slope of the equation: $y = \dfrac{2}{3}x + \dfrac{1}{3}$

7. Write the equation given the: slope = -7, and one set of points: $(3,-2)$

8. Find the slope using the formula: $m = \dfrac{rise}{run}$

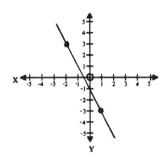

9. Describe the slopes as: <u>positive or negative</u> and <u>greater than one or less than one</u>.

a. _____

b. _____

188

Cumulative Review 10

1. Solve by addition: $2x - y = 5$
   $x + 2y = 5$

2. Solve by substitution:
   $3x + y = 6$
   $x - 2y = 2$

3. Find the LCD: $\dfrac{5}{6x^2} + \dfrac{7}{2x} - \dfrac{1}{5x^3}$

4. Solve: $\dfrac{x-3}{x+8} = \dfrac{x-6}{x+2}$

5. Solve for x: $x^2 - 6x = 16$

6. Simplify: $\dfrac{x^2 - 9x + 14}{3x^2 - 21x}$

7. Factor: $x^4 - 81$

8. Factor: $9x^2 - 25$

9. Solve and put on the number line:
   $7x < -35$

   ———————|———————
                 0

10. Solve: $x - 12 = 30$

11. Solve: $-6x = 21$

12. Simplify: $\dfrac{x^{-1}}{x^{-5}}$

13. Divide: $3x + 7 \overline{\smash{\big)}\ 6x^2 + 29x + 35}$

14. Simplify: $7 - (3 + 2\{7 - 2\} - 14) - 9$

Chapter 13
Radicals

What is a Radical? A radical is a symbol: $\sqrt{}$. This symbol means "the square root of" which means "what number when multiplied by itself gives you ___?" For example, $\sqrt{81}$ means "what number when multiplied by itself gives you 81?" The answer is 9, therefore, 9 is the <u>square root</u> of 81. Since 81 is the product of a number times itself, 81 is a **perfect square**.

Examples
Find the square root of:

a. $\sqrt{64} = 8$
because $(8)(8) = 64$

b. $\sqrt{a^8} = a^4$
because $a^4 \cdot a^4 = a^8$

c. $\sqrt{144x^6} = 12x^3$
because $12x^3 \cdot 12x^3 = 144x^6$

Name the square root of:

1. $\sqrt{100}$

2. $\sqrt{9}$

3. $\sqrt{x^{10}}$

4. $\sqrt{a^{18}}$

5. $\sqrt{25c^8}$

6. $\sqrt{36x^{16}}$

7. $\sqrt{4}$

8. $\sqrt{y^2}$

9. $\sqrt{9z^4}$

10. $\sqrt{b^{14}x^{10}}$

11. $\sqrt{16c^6}$

12. $\sqrt{64a^{12}b^6}$

13. $\sqrt{225}$

14. $\sqrt{a^6}$

15. $\sqrt{25z^{22}a^4}$

16. $\sqrt{c^6b^4}$

17. $\sqrt{81z^{18}a^{14}}$

18. $\sqrt{x^{12}}$

19. $\sqrt{121}$

20. $\sqrt{169z^8}$

21. $\sqrt{225b^{10}}$

22. $\sqrt{9x^{12}}$

23. $\sqrt{1}$

24. $\sqrt{49}$

Simplifying Radicals - if the term under the radical is not a perfect square, try to:

1. divide a perfect square into the term
2. rewrite the problem as the <u>factors</u> of that term, then
3. bring the perfect square out from under the radical

Examples

a. $\sqrt{50}$
$= \sqrt{25 \cdot 2}$
$= 5\sqrt{2}$

b. $\sqrt{x^7}$
$= \sqrt{x^6 \cdot x}$
$= x^3 \sqrt{x}$

c. $\sqrt{52y^{13}}$
$= \sqrt{4 \cdot 13 \cdot y^{12} \cdot y}$
$= 2y^6 \sqrt{13y}$

Note the perfect squares, ($25, x^6, 4,$ and y^{12}), that divide into the examples.

Simplify

1. $\sqrt{28}$
2. $\sqrt{44}$
3. $\sqrt{45}$
4. $\sqrt{63}$
5. $\sqrt{a^3}$

6. $\sqrt{b^{21}}$
7. $\sqrt{x^{17}}$
8. $\sqrt{c^{15}}$
9. $\sqrt{90a^{11}}$
10. $\sqrt{32b^3}$

11. $\sqrt{52c^{21}}$
12. $\sqrt{12x^{19}}$
13. $\sqrt{8}$
14. $\sqrt{z^{19}}$
15. $\sqrt{40x^{23}}$

16. $\sqrt{27a^{17}}$
17. $\sqrt{z^5}$
18. $\sqrt{20}$
19. $\sqrt{a^{25}}$
20. $\sqrt{12b^9}$

Adding and Subtracting Radicals - you can combine (add or subtract) <u>like radicals only</u>. Add or subtract the <u>numerical coefficients</u> of the like radicals; do not add or subtract the numbers under the radicals.

Examples

a. $4\sqrt{3} + 2\sqrt{3} = 6\sqrt{3}$ b. $5\sqrt{6} + 2\sqrt{7} - 3\sqrt{6} + 5\sqrt{7} = 2\sqrt{6} + 7\sqrt{7}$

Simplify

1. $3\sqrt{2} + 5\sqrt{2}$
2. $4\sqrt{7} - 1\sqrt{7}$
3. $6\sqrt{3} - 4\sqrt{3}$
4. $8\sqrt{10} + 4\sqrt{10}$
5. $4\sqrt{14} + 2\sqrt{14}$
6. $6\sqrt{7} - 3\sqrt{7}$
7. $6\sqrt{11} + 2\sqrt{11}$
8. $4\sqrt{10} - 3\sqrt{10}$
9. $4\sqrt{13} + 5\sqrt{13}$

10. $4\sqrt{2} + 3\sqrt{5} + 2\sqrt{2}$
11. $8\sqrt{6} - 2\sqrt{5} + 3\sqrt{6}$
12. $4\sqrt{3} + 3\sqrt{7} + 2\sqrt{3}$
13. $3\sqrt{5} - 4\sqrt{5} + 7\sqrt{6}$
14. $4\sqrt{10} + 2\sqrt{10} + 4\sqrt{2}$
15. $8\sqrt{7} + 2\sqrt{7} - 3\sqrt{2}$
16. $3\sqrt{6} + 4\sqrt{6} + 9\sqrt{3}$
17. $8\sqrt{11} + 4\sqrt{2} - 3\sqrt{2}$
18. $4\sqrt{3} - 7\sqrt{10} + 1\sqrt{10}$

19. $4\sqrt{3} + 7\sqrt{6} - 4\sqrt{3} + 2\sqrt{6}$
20. $3\sqrt{2} + 9\sqrt{7} - 4\sqrt{3} + 2\sqrt{7}$
21. $2\sqrt{8} + 1\sqrt{3} - \sqrt{8} + 2\sqrt{3}$
22. $3\sqrt{6} - 5\sqrt{5} + \sqrt{5} - 2\sqrt{6}$
23. $\sqrt{3} + 7\sqrt{2} - \sqrt{3} - 6\sqrt{2}$
24. $\sqrt{6} + 3\sqrt{7} - 3\sqrt{7} - 2\sqrt{6}$
25. $4\sqrt{2} + 7\sqrt{2} - 4\sqrt{6} + 2\sqrt{6}$
26. $9\sqrt{5} - 3\sqrt{5} - 11\sqrt{3} + 7\sqrt{3}$
27. $\sqrt{6} + 4\sqrt{2} - 7\sqrt{2} + \sqrt{6}$

Multiplicaation of Radicals - a radical can be multiplied by a radical and the product will be a radical. A radical <u>cannot</u> be multiplied by a rational number. Often the product of two radicals can be simplified.   **Examples**

a. $\sqrt{3} \cdot \sqrt{5}$
   $= \sqrt{15}$

b. $\sqrt{3} \cdot 4\sqrt{6}$
   $= 4\sqrt{18}$
   $= 4\sqrt{9 \cdot 2}$
   $= 4(3)\sqrt{2}$
   $= 12\sqrt{2}$

c. $2\sqrt{5x} \cdot 5\sqrt{10x^4}$
   $= 10\sqrt{50x^5}$
   $= 10\sqrt{25 \cdot 2 \cdot x^4 \cdot x}$
   $= 10(5)x^2\sqrt{2x}$
   $= 50x^2\sqrt{2x}$

Multiply and simplify if needed

1. $\sqrt{2} \cdot \sqrt{11}$

2. $\sqrt{3} \cdot \sqrt{10}$

3. $\sqrt{5} \cdot \sqrt{7}$

4. $\sqrt{3} \cdot \sqrt{13}$

5. $\sqrt{7} \cdot \sqrt{3}$

6. $\sqrt{11} \cdot \sqrt{5}$

7. $6\sqrt{14} \cdot \sqrt{3}$

8. $3\sqrt{7} \cdot \sqrt{2}$

9. $7\sqrt{2} \cdot 3\sqrt{3}$

10. $9\sqrt{3} \cdot 5\sqrt{5}$

11. $\sqrt{2} \cdot \sqrt{6}$

12. $\sqrt{6} \cdot \sqrt{8}$

13. $\sqrt{5} \cdot 3\sqrt{8}$

14. $4\sqrt{3} \cdot 5\sqrt{6}$

15. $4\sqrt{12} \cdot 2\sqrt{2}$

16. $7\sqrt{2} \cdot 2\sqrt{8}$

17. $2\sqrt{12x} \cdot 3\sqrt{2x}$

18. $3\sqrt{10x^2} \cdot 5\sqrt{2x^3}$

19. $5\sqrt{2x^4} \cdot 2\sqrt{6x^3}$

20. $3\sqrt{6x} \cdot 2\sqrt{10x^4}$

More Multiplication of Radicals - when multiplying a radical and a <u>binomial</u> that contains a radical, use the distributive property.(multiply both terms inside the parentheses by the term outside the parentheses) <u>Often the radicals can be simplified</u>.

Examples

a. $\sqrt{3}(\sqrt{2} - \sqrt{7})$
  $= \sqrt{6} - \sqrt{21}$

b. $3\sqrt{5}(\sqrt{7} - 6)$
  $= 3\sqrt{35} - 18\sqrt{5}$

c. $4\sqrt{2}(3\sqrt{8} + 2\sqrt{10})$
  $= 12\sqrt{16} + 8\sqrt{20}$
  $= 12(4) + 8\sqrt{4 \cdot 5}$
  $= 48 + 8(2)\sqrt{5}$
  $= 48 + 16\sqrt{5}$

Multiply and simplify if needed

1. $\sqrt{5}(\sqrt{3} - \sqrt{2})$

2. $\sqrt{2}(\sqrt{5} + \sqrt{7})$

3. $\sqrt{3}(\sqrt{7} - \sqrt{5})$

4. $2\sqrt{11}(\sqrt{5} + \sqrt{6})$

5. $2\sqrt{5}(\sqrt{7} - 4)$

6. $4\sqrt{2}(2\sqrt{3} + \sqrt{5})$

7. $3\sqrt{7}(5\sqrt{5} - 7\sqrt{3})$

8. $3\sqrt{10}(4\sqrt{3} + \sqrt{7})$

9. $3\sqrt{6}(2\sqrt{3} + 3\sqrt{2})$

10. $7\sqrt{2}(5\sqrt{6} - 3\sqrt{9})$

11. $4\sqrt{3}(2\sqrt{12} - 5\sqrt{8})$

12. $5\sqrt{8}(3\sqrt{2} - 6)$

13. $4\sqrt{6}(7 + 6\sqrt{8})$

14. $4\sqrt{10}(2\sqrt{2} - 5\sqrt{5})$

15. $3\sqrt{10}(4\sqrt{2} + 3\sqrt{10})$

16. $6\sqrt{5}(2\sqrt{10} + 3\sqrt{5})$

17. $2\sqrt{3}(4\sqrt{6} - 2\sqrt{8})$

18. $4\sqrt{8}(3\sqrt{3} - 2\sqrt{5})$

More Multiplication of Radicals - if a radical is multiplied by "itself", the product is "itself" without the radical. For example, $\sqrt{5} \cdot \sqrt{5} = \sqrt{25} = 5$. Notice the middle step, $\sqrt{25}$, can be left out.

The CONJUGATE of a binomial - is another binomial that has the same terms but the opposite signs between the terms. The conjugate of $(\sqrt{6} - 2\sqrt{5})$ is $(\sqrt{6} + 2\sqrt{5})$.

When multiplying conjugates - the "OI" portion of FOIL will cancel out each other, therefore, it is only necessary to multiply the "FL" portion of FOIL.

Examples

a. $(3 - \sqrt{2})(3 + \sqrt{2})$
    = 9 - 2
    = 7
    (note: $\sqrt{2} \times \sqrt{2} = 2$)

b. $(4\sqrt{2} + 3\sqrt{5})(4\sqrt{2} - 3\sqrt{5})$
    = 16(2) - 9(5)
    = 32 - 45
    = -13

Multiply

1. $(2 + \sqrt{5})(2 - \sqrt{5})$
2. $(4 - \sqrt{3})(4 + \sqrt{3})$
3. $(\sqrt{5} + 2)(\sqrt{5} - 2)$
4. $(3\sqrt{2} + 5)(3\sqrt{2} - 5)$
5. $(4\sqrt{3} - 6)(4\sqrt{3} + 6)$
6. $(5\sqrt{2} - \sqrt{3})(5\sqrt{2} + \sqrt{3})$
7. $(4\sqrt{3} + 2)(4\sqrt{3} - 2)$
8. $(3\sqrt{5} - 3\sqrt{3})(3\sqrt{5} + 3\sqrt{3})$
9. $(6\sqrt{6} + 2\sqrt{3})(6\sqrt{6} - 2\sqrt{3})$
10. $(4\sqrt{7} - 5\sqrt{2})(4\sqrt{7} + 5\sqrt{2})$
11. $(6\sqrt{3} - 5)(6\sqrt{3} + 5)$
12. $(\sqrt{2} + 5\sqrt{7})(\sqrt{2} - 5\sqrt{7})$
13. $(8\sqrt{2} + 3)(8\sqrt{2} - 3)$
14. $(6\sqrt{2} - \sqrt{3})(6\sqrt{2} + \sqrt{3})$
15. $(\sqrt{10} - 1)(\sqrt{10} + 1)$

More Multiplication of Radicals - when multiplying ordinary binomials (not conjugates) the FOIL method is used.

Examples

a. $(4 - \sqrt{3})(2 + \sqrt{3})$
 $= 8 + 4\sqrt{3} - 2\sqrt{3} - 3)$
 $= 5 + 2\sqrt{3}$

b. $(2\sqrt{5} + 4)(3\sqrt{5} + 1)$
 $= 6(5) + 2\sqrt{5} + 12\sqrt{5} + 4$
 $= 30 + 14\sqrt{5} + 4$
 $= 34 + 14\sqrt{5}$

Multiply

1. $(5 - \sqrt{2})(3 + \sqrt{2})$

2. $(6 - \sqrt{3})(2 - \sqrt{3})$

3. $(2 + \sqrt{6})(3 - \sqrt{6})$

4. $(3\sqrt{2} + 1)(2\sqrt{2} + 5)$

5. $(4\sqrt{5} - 6)(2\sqrt{5} + 3)$

6. $(6 + 2\sqrt{3})(5 - \sqrt{3})$

7. $(2 - 5\sqrt{6})(3 - 2\sqrt{6})$

8. $(3\sqrt{7} - \sqrt{6})(\sqrt{7} + 4\sqrt{6})$

9. $(4\sqrt{2} + 2\sqrt{3})(\sqrt{2} - \sqrt{3})$

10. $(\sqrt{2} - 4\sqrt{7})(3\sqrt{2} + 2\sqrt{7})$

11. $(\sqrt{5} - 9\sqrt{2})(6\sqrt{5} + \sqrt{2})$

12. $(2\sqrt{3} - \sqrt{5})(\sqrt{3} + 4\sqrt{5})$

13. $(1 + 4\sqrt{2})(4 - 2\sqrt{2})$

14. $(4 - \sqrt{11})(3 - \sqrt{11})$

15. $(10 + 2\sqrt{2})(2 + 3\sqrt{2})$

16. $(4\sqrt{3} + 7)(5\sqrt{3} - 6)$

17. $(\sqrt{2} + 7)(\sqrt{2} - 3)$

18. $(4 + 3\sqrt{7})(1 + 2\sqrt{7})$

Simplifying Radicals in Fraction Form- if there is a fraction under a radical, $\sqrt{\frac{2}{3}}$, each term under the radical can be put under its own radical, $\frac{\sqrt{2}}{\sqrt{3}}$. The reverse is also true. If both terms of a fraction are under separate radicals, both terms can be put under the same radical.

Dividing Radicals - it is permissable to divide radicals into radicals, but it is not permissable to divide <u>rational numbers</u> into radicals. If the terms of the fraction divide evenly, divide them, then check the radical to see if it can be further simplified.

Examples

a. $\sqrt{\frac{12}{6}}$
$= \sqrt{2}$

b. $\frac{\sqrt{8}}{\sqrt{2}}$
$= \sqrt{4}$
$= 2$

c. $\frac{10\sqrt{24}}{5\sqrt{3}}$
$= 2\sqrt{8}$
$= 2\sqrt{4 \cdot 2}$
$= 2(2)\sqrt{2}$
$= 4\sqrt{2}$

Simplify by dividing

1. $\frac{\sqrt{6}}{\sqrt{2}}$

2. $\sqrt{\frac{10}{5}}$

3. $\sqrt{\frac{14}{7}}$

4. $\frac{\sqrt{90}}{\sqrt{2}}$

5. $\frac{\sqrt{48}}{\sqrt{2}}$

6. $\sqrt{\frac{120}{3}}$

7. $\frac{\sqrt{27}}{\sqrt{3}}$

8. $\sqrt{\frac{10}{2}}$

9. $\frac{6\sqrt{54}}{2\sqrt{3}}$

10. $\frac{9\sqrt{84}}{3\sqrt{3}}$

11. $\frac{\sqrt{50}}{\sqrt{2}}$

12. $\frac{\sqrt{48}}{\sqrt{4}}$

13. $\frac{20\sqrt{36}}{4\sqrt{3}}$

14. $\sqrt{\frac{24}{3}}$

15. $\frac{16\sqrt{45}}{2\sqrt{9}}$

16. $\frac{20\sqrt{36}}{4\sqrt{2}}$

Simplifying Radicals That Are Perfect Squares - if there is a fraction inside a radical, the numerator and denominator can each be put under their own radical. For example, $\sqrt{\frac{5}{8}}$ can be written $\frac{\sqrt{5}}{\sqrt{8}}$.

If the terms under the radicals are perfect squares, or if they can be reduced to terms that are perfect squares, then can be brought out from under the radical.

Examples

a. $\sqrt{\frac{25}{4}} = \frac{\sqrt{25}}{\sqrt{4}} = \frac{5}{2}$
b. $\frac{\sqrt{18}}{\sqrt{200}} = \sqrt{\frac{18}{200}} = \sqrt{\frac{9}{100}} = \frac{3}{10}$

Simplify

1. $\sqrt{\frac{25}{9}}$

2. $\frac{\sqrt{9}}{\sqrt{49}}$

3. $\sqrt{\frac{16}{9}}$

4. $\frac{\sqrt{2}}{\sqrt{128}}$

5. $\sqrt{\frac{18}{8}}$

6. $\sqrt{\frac{3}{12}}$

7. $\frac{\sqrt{a^2}}{\sqrt{b^2}}$

8. $\sqrt{\frac{x^6}{y^2}}$

9. $\sqrt{\frac{4r^2}{9a^4}}$

10. $\sqrt{\frac{36y^2}{25b^8}}$

11. $\frac{\sqrt{64}}{\sqrt{25}}$

12. $\sqrt{\frac{50}{18}}$

13. $\frac{\sqrt{3}}{\sqrt{75}}$

14. $\sqrt{\frac{x^6}{9y^6}}$

15. $\sqrt{\frac{48}{27}}$

16. $\frac{\sqrt{125}}{\sqrt{45a^4}}$

17. $\sqrt{\frac{4y^6}{9}}$

18. $\frac{\sqrt{100}}{\sqrt{49}}$

19. $\sqrt{\frac{128}{50}}$

20. $\sqrt{\frac{49x^2}{a^2b^2}}$

21. $\frac{\sqrt{49}}{\sqrt{64}}$

Simplifying Radicals by Rationalizing the Denominator - to change an "irrational denominator" to a "rational denominator", multiply the denominator and its numerator by the denominator, (or any other number that will give a product that is a perfect square.) This is called **rationalizing the denominator**.

Examples

a. $\dfrac{\sqrt{x}}{\sqrt{3}} = \dfrac{\sqrt{x}}{\sqrt{3}} \cdot \dfrac{\sqrt{3}}{\sqrt{3}} = \dfrac{\sqrt{3x}}{3}$ b. $\sqrt{\dfrac{3}{8}} = \dfrac{\sqrt{3}}{\sqrt{8}} \cdot \dfrac{\sqrt{2}}{\sqrt{2}} = \dfrac{\sqrt{6}}{\sqrt{16}} = \dfrac{\sqrt{6}}{4}$

Rationalize the denominator

1. $\sqrt{\dfrac{1}{2}}$

2. $\dfrac{\sqrt{3}}{\sqrt{7}}$

3. $\sqrt{\dfrac{a}{6}}$

4. $\dfrac{\sqrt{x}}{\sqrt{11}}$

5. $\dfrac{\sqrt{5}}{\sqrt{8}}$

6. $\sqrt{\dfrac{11}{50}}$

7. $\dfrac{\sqrt{13}}{\sqrt{20b^7}}$

8. $\dfrac{\sqrt{3}}{\sqrt{x^3}}$

9. $\sqrt{\dfrac{3}{20}}$

10. $\dfrac{\sqrt{1}}{\sqrt{5}}$

11. $\dfrac{\sqrt{x}}{\sqrt{7}}$

12. $\sqrt{\dfrac{3}{10}}$

13. $\dfrac{\sqrt{7}}{\sqrt{20}}$

14. $\sqrt{\dfrac{3}{5}}$

15. $\sqrt{\dfrac{3x}{7}}$

16. $\dfrac{\sqrt{5}}{\sqrt{18}}$

More Simplifying Radicals by Rationalizing the Denominator - if the denominator is a binomial containing a radical, to rationalize the denominator, **multiply** the denominator and numerator **by the conjugate of the denominator**.

Examples

a. $\dfrac{3}{\sqrt{5}-2} = \dfrac{3}{\sqrt{5}-2} \cdot \dfrac{\sqrt{5}+2}{\sqrt{5}+2} = \dfrac{3\sqrt{5}+6}{5-4} = \dfrac{3\sqrt{5}+6}{1} = 3\sqrt{5}+6$

b. $\dfrac{\sqrt{6}}{4-\sqrt{3}} = \dfrac{\sqrt{6}}{4-\sqrt{3}} \cdot \dfrac{4+\sqrt{3}}{4+\sqrt{3}} = \dfrac{4\sqrt{6}+\sqrt{18}}{16-3} = \dfrac{4\sqrt{6}+\sqrt{9\cdot 2}}{13} = \dfrac{4\sqrt{6}+3\sqrt{2}}{13}$

Rationalize the denominator

1. $\dfrac{1}{\sqrt{2}+3}$

2. $\dfrac{6}{\sqrt{5}+2}$

3. $\dfrac{10}{4-\sqrt{3}}$

4. $\dfrac{-7}{5-3\sqrt{3}}$

5. $\dfrac{9}{5\sqrt{7}+3}$

6. $\dfrac{\sqrt{6}}{2\sqrt{3}+1}$

7. $\dfrac{\sqrt{3}}{4\sqrt{2}-1}$

Simplifying Radicals Continued

8. $\dfrac{2\sqrt{5}}{2\sqrt{2}+\sqrt{6}}$

9. $\dfrac{\sqrt{7}}{\sqrt{3}+\sqrt{2}}$

10. $\dfrac{6}{\sqrt{2}-5}$

11. $\dfrac{5}{2-\sqrt{5}}$

12. $\dfrac{\sqrt{6}}{3\sqrt{2}+\sqrt{10}}$

13. $\dfrac{9}{6-\sqrt{5}}$

14. $\dfrac{-1}{5+\sqrt{6}}$

15. $\dfrac{\sqrt{3}}{4\sqrt{6}-1}$

16. $\dfrac{-4}{\sqrt{7}+10}$

17. $\dfrac{\sqrt{11}}{\sqrt{3}+2}$

Simplifying and Combining Radicals - when adding or subtracting radicals, they must be broken down and made to be "like radicals" before they can be combined.

Examples

a. $2\sqrt{18} + \sqrt{8}$
 $= 2\sqrt{9 \cdot 2} + \sqrt{4 \cdot 2}$
 $= 2(3)\sqrt{2} + 2\sqrt{2}$
 $= 6\sqrt{2} + 2\sqrt{2}$
 $= 8\sqrt{2}$

b. $2\sqrt{12} + 3\sqrt{108} - \sqrt{48}$
 $= 2\sqrt{4 \cdot 3} + 3\sqrt{36 \cdot 3} - \sqrt{16 \cdot 3}$
 $= 2(2)\sqrt{3} + 3(6)\sqrt{3} - 4\sqrt{3}$
 $= 4\sqrt{3} + 18\sqrt{3} - 4\sqrt{3}$
 $= 18\sqrt{3}$

Simplify and combine

1. $\sqrt{18} + \sqrt{32}$

2. $2\sqrt{45} - \sqrt{80}$

3. $\sqrt{75} + 4\sqrt{27}$

4. $5\sqrt{300} - 4\sqrt{12}$

5. $\sqrt{72} - \sqrt{8} + 2\sqrt{50}$

6. $\sqrt{18} + \sqrt{50} - 3\sqrt{98}$

7. $\sqrt{8} - 2\sqrt{200} + \sqrt{50}$

8. $\sqrt{80} + 2\sqrt{45} + \sqrt{20}$

More Simplifying and Combining Radicals - the term $\frac{\sqrt{5}}{2}$ can be written $\frac{1}{2}\sqrt{5}$. Knowing this is helpful when simplifying and combining radicals containing fractions where "rationalizing the denominator" is necessary.

Example
$$\sqrt{45} + \sqrt{\tfrac{1}{5}}$$
$$= \sqrt{9 \cdot 5} + \frac{\sqrt{1}}{\sqrt{5}} \cdot \frac{\sqrt{5}}{\sqrt{5}}$$
$$= 3\sqrt{5} + \frac{\sqrt{5}}{5}$$
$$= 3\sqrt{5} + \tfrac{1}{5}\sqrt{5}$$
$$= 3\tfrac{1}{5}\sqrt{5}$$

Simplify and combine

1. $\sqrt{18} + \dfrac{\sqrt{1}}{\sqrt{2}}$

2. $\dfrac{\sqrt{1}}{\sqrt{3}} + \sqrt{27}$

3. $\sqrt{24} + \dfrac{\sqrt{1}}{\sqrt{6}}$

4. $\sqrt{40} - \sqrt{\tfrac{2}{5}}$

5. $\sqrt{24} + 6\sqrt{\tfrac{2}{3}}$

6. $3\sqrt{10} - 5\sqrt{\tfrac{2}{5}}$

7. $\sqrt{42} + 7\sqrt{\tfrac{6}{7}}$

8. $2\sqrt{21} - 21\sqrt{\tfrac{3}{7}}$

9. $\sqrt{\tfrac{3}{5}} + \sqrt{60} + 9\sqrt{\tfrac{5}{3}}$

10. $\sqrt{32} + \sqrt{\tfrac{1}{2}} - \sqrt{2}$

11. $\sqrt{24} + \sqrt{\tfrac{3}{2}} - 6\sqrt{\tfrac{2}{3}}$

12. $15\sqrt{\tfrac{2}{5}} - \sqrt{40} + 8\sqrt{\tfrac{5}{2}}$

Radical Equations - to solve, move all terms except the radical to the other side of the equation then <u>square the radical **and** the other side of the equation.</u> This gets rid of the radical. Proceed to solve the equation.

Examples

a. $\sqrt{x} = 7$
$(\sqrt{x})^2 = (7)^2$
$x = 49$

b. $\sqrt{x+3} = 5$
$(\sqrt{x+3})^2 = (5)^2$
$x + 3 = 25$
$\quad -3 \quad -3$
$x = 22$

c. $\sqrt{x} + 6 = 9$
$\quad\quad -6 \quad -6$
$\sqrt{x} = 3$
$(\sqrt{x})^2 = (3)^2$
$x = 9$

Solve

1. $\sqrt{x} = 6$

2. $\sqrt{x} = 9$

3. $\sqrt{x} = 1$

4. $\sqrt{x+2} = 5$

5. $\sqrt{x-5} = 4$

6. $\sqrt{x+8} = 2$

7. $\sqrt{x} + 7 = 10$

8. $\sqrt{x} + 10 = 15$

9. $\sqrt{x} - 3 = 2$

10. $\sqrt{x-8} = 10$

11. $\sqrt{x} + 5 = 11$

12. $\sqrt{x} = 12$

13. $\sqrt{x} - 7 = 11$

14. $\sqrt{x-2} = 7$

15. $\sqrt{x} - 6 = 4$

16. $\sqrt{x} = 3$

More Radical Equations - be sure to move all terms except the radical to the other side of the equation before "squaring" both sides of the equation.

Examples

a. $3\sqrt{x} = 12$
$\frac{3\sqrt{x}}{3} = \frac{12}{3}$
$\sqrt{x} = 4$
$(\sqrt{x})^2 = (4)^2$
$x = 16$

b. $\sqrt{2x+3} = 4$
$(\sqrt{2x+3})^2 = (4)^2$
$2x+3 = 16$
$-3 -3$
$\frac{2x}{2} = \frac{13}{2}$
$x = \frac{13}{2}$

c. $3\sqrt{x+2} + 6 = 9$
$\phantom{3\sqrt{x+2}}-6 -6$
$\frac{3\sqrt{x+2}}{3} = \frac{3}{3}$
$\sqrt{x+2} = 1$
$(\sqrt{x+2})^2 = (1)^2$
$x+2 = 1$
$-2 -2$
$x = -1$

Solve

1. $4\sqrt{x} = 8$

2. $8\sqrt{x} = 56$

3. $\sqrt{2x+5} = 7$

4. $\sqrt{3x+1} = 5$

5. $11\sqrt{x+2} = 33$

6. $7\sqrt{x-5} = 35$

7. $4\sqrt{x} + 2 = 10$

8. $2\sqrt{x} - 4 = 10$

9. $7\sqrt{x-1} + 3 = 24$

10. $3\sqrt{x+2} + 6 = 18$

11. $\sqrt{2x+5} = 3$

12. $7\sqrt{x} = 28$

13. $3\sqrt{2x-1} - 4 = 11$

Review of Radicals

Simplify, add, subtract, multiply, divide, rationalize the denominator, or solve:

1. $\sqrt{25}$

2. $\sqrt{45x^{15}}$

3. $\sqrt{8x^2} \cdot 2\sqrt{3x^4}$

4. $5\sqrt{3}(2\sqrt{6}-4)$

5. $(\sqrt{10}-1)(\sqrt{10}+4)$

6. $\sqrt{\frac{x}{10}}$

7. $7\sqrt{6}+\sqrt{6}$

8. $\dfrac{\sqrt{2}}{\sqrt{3x}}$

9. $\dfrac{2\sqrt{3}}{\sqrt{7}-5\sqrt{2}}$

10. $\sqrt{x-9}=3$

11. $\sqrt{54}+6\sqrt{\frac{2}{3}}$

12. $\sqrt{a^{26}}$

13. $\sqrt{10} \cdot \sqrt{11}$

14. $\sqrt{\frac{9}{100}}$

15. $\sqrt{13}(\sqrt{2}-4\sqrt{3})$

16. $(3+2\sqrt{6})(3-2\sqrt{6})$

17. $8\sqrt{2}-7\sqrt{2}$

18. $\sqrt{\frac{20}{5}}$

19. $(2\sqrt{5}+3)(\sqrt{5}+2)$

20. $3\sqrt{20}-3\sqrt{45}$

21. $3\sqrt{x-1}+7=16$

Cumulative Review 11

1. Find the slope using: $m = \frac{y_1 - y_2}{x_1 - x_2}$
   (5,-3), (-2, 4)

2. Write the equation given slope = -3 and (4, -2)

3. Write the equations that are:

   equal to:

   parallel to:

   perpendicular to $2x + 5y = -1$

4. Solve by graphing: $x - y = 2$
   $2x + y = -8$

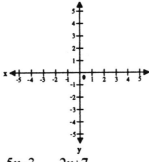

5. Simplify: $\frac{5x-3}{x-1} + \frac{2x+7}{1-x}$

6. Simplify: $\frac{x^2 + 7x + 6}{x^2 - 1}$

7. Factor: $4x^2y - 6xy^2$

8. Solve, put on the number line, and give the solution set: $2 < x - 1 > -3$

   ———•——— $x = \{\quad\}$

9. Solve: $3x - 1 = 19$

10. Simplify: $\left(\frac{x^{-3}}{2x^2}\right)^{-1}$

11. Multiply: $(3x + 7)(x - 4)$

12. Simplify: $4(3 - 2\{6 + 3\} + 13) + 9$

Chapter 14
Quadratic Equations

Factoring - in chapter 9, solving quadratic equations by factoring was covered. A short review of that material is appropriate for this chapter.

Examples

a. $2x(x - 4) = 0$
   $2x = 0$ or $x - 4 = 0$
   $x = 0$ or $x = 4$

b. $(3x - 1)(2x + 5) = 0$
   $3x - 1 = 0$ or $2x + 5 = 0$
   $+1 +1 -5 = -5$
   $\frac{3x}{3} = \frac{1}{3} \frac{2x}{2} = \frac{-5}{2}$
   $x = \frac{1}{3} \; ; \; x = -\frac{5}{2}$

c. $x^2 - 6x = +16$
   $ -16 -16$
   $x^2 - 6x - 16 = 0$
   $(x - 8)(x + 2) = 0$
   $x = +8 \; ; \; x = -2$

Solve by Factoring

1. $(2x + 1)(4x - 3) = 0$

2. $9x^2 - 27x = 0$

3. $x^2 + 4x - 12 = 0$

4. $x(x - 9) = 0$

5. $x^2 - 16 = 9$

6. $4x^2 - 7x = -3$

7. $(a)(b) = 0$

8. $(x - 1)(x + 12) = 0$

9. $4x^2 - 1 = 0$

10. $5x(x + 6) = 0$

11. $x^2 = -13x - 30$

12. $5x^2 + 20x = 0$

Quadratic Formula - a second way to solve quadratic equations is to use a <u>formula</u>. The formula to use for <u>all quadratic equations</u> is:

$$x = \frac{-b \pm \sqrt{b^2 - 4ac}}{2a}$$

There can be as many as 10 steps involved in determining the values of x when using this formula.

Example

Solve by using the quadratic formula: $x^2 - 10x + 21 = 0$

1. $x^2 - 10x + 21 = 0$ The problem is written down.

2. $x = \dfrac{-b \pm \sqrt{b^2 - 4ac}}{2a}$ The formula is written down.

3. a = 1 (the number in front of x^2)   The values of a, b, and c are determined.
   b = -10 (the number in front of x)
   c = +21 (the constant - the third term)

4. $x = \dfrac{-(-10) \pm \sqrt{(10)^2 - 4(1)(21)}}{2(1)}$ The values are substituted into the formula.

5. $x = \dfrac{10 \pm \sqrt{100 - 84}}{2}$ The parentheses are removed.

6. $x = \dfrac{10 \pm \sqrt{16}}{2}$ The integers under the radical are combined.

7. $x = \dfrac{10 \pm 4}{2}$ The square root of the radical is taken.

8. $x = \dfrac{10+4}{2}$; $x = \dfrac{10-4}{2}$ X is equated to the two values.

9. $x = \dfrac{14}{2}$; $x = \dfrac{6}{2}$ The numerators are added.

10. x = 7 ; x = 3   The fractions are simplified.

Solve by using the quadratic equation.

1. $x^2 + 8x + 15 = 0$

   $x = \dfrac{-b \pm \sqrt{b^2 - 4ac}}{2a}$

2. $x^2 - 11x + 30 = 0$

   $x = \dfrac{-b \pm \sqrt{b^2 - 4ac}}{2a}$

3. $x^2 + 5x - 24 = 0$

   $x = \dfrac{-b \pm \sqrt{b^2 - 4ac}}{2a}$

4. $x^2 - 3x - 18 = 0$

   $x = \dfrac{-b \pm \sqrt{b^2 - 4ac}}{2a}$

5. $4x^2 - 12x + 5 = 0$

   $x = \dfrac{-b \pm \sqrt{b^2 - 4ac}}{2a}$

6. $2x^2 - 9x - 5 = 0$

   $x = \dfrac{-b \pm \sqrt{b^2 - 4ac}}{2a}$

Completing the Binomial Square - this is the third method for solving quadratic equations. A trinomial is changed so that it will factor as a binomial square, then the square root of the binomial square is taken. There are three steps, and in the first step the constant is moved to the other side of the equation.

Example of the First Step

a. $x^2 - 8x + 7 = 0$
   $ -7 -7$
   $x^2 - 8x + \_\_ = -7$

b. $x^2 - 9x - 12 = 10$
   $ +12 +12$
   $x^2 - 9x + \_\_ = +22$

Move the constant to the other side of the equation

1. $x^2 - 2x - 15 = 0$

2. $x^2 + 8x - 9 = 0$

3. $x^2 + 4x - 12 = 0$

4. $x^2 + 10x + 21 = 0$

5. $x^2 - 4x + 3 = 0$

6. $x^2 + 7x - 14 = 4$

7. $x^2 - 3x - 20 = 20$

8. $x^2 - 7x + 2 = -8$

9. $x^2 - 5x - 11 = 3$

10. $x^2 - 2x - 24 = 0$

11. $x^2 + 8x - 20 = 0$

12. $x^2 + 12x + 27 = 0$

13. $x^2 - 5x - 14 = 0$

14. $x^2 - x - 20 = 0$

15. $x^2 + 11x + 30 = 0$

16. $x^2 + 7x + 3 = -7$

17. $x^2 - 2x - 10 = 0$

18. $x^2 - 10x + 25 = 0$

Completing the Binomial Square Continued - in the second step, the blank is filled with an integer that makes the trinomial factor as a **binomial square**.
Follow these steps:
1) Put ( )^2 on the next line. **2)** Take half of the middle term, put the number in the parentheses, square that number, and put that number in the blank above. Then put that number on the other side of the equation. **3)** Put an "x" in the parentheses and add the numbers on the other side of the equation.

Examples of Step Two

a. $x^2 - 8x + (16) = -7 + (16)$
   $(x - 4)^2 = 9$

b. $x^2 + 9x + (\frac{81}{4}) = 22 + (\frac{81}{4})$
   $(x + \frac{9}{2})^2 = \frac{169}{4}$

$\frac{22}{1} = \frac{88}{4}$
$\frac{81}{4} = \frac{81}{4}$
$\phantom{\frac{81}{4}} \frac{169}{4}$

Work the second step of completing the binomial square.

1. $x^2 - 2x + \underline{} = 15$

2. $x^2 + 8x + \underline{} = -7$

3. $x^2 - 6x + \underline{} = 7$

4. $x^2 + 4x + \underline{} = 12$

5. $x^2 + 10x + \underline{} = -21$

6. $x^2 - 4x + \underline{} = -3$

7. $x^2 + 7x + \underline{} = 18$

8. $x^2 - 3x + \underline{} = 40$

9. $x^2 - 5x + \underline{} = 14$

10. $x^2 + 11x + \underline{} = 26$

Completing the Binomial Square Continued - in the third step of completing the binomial square, follow these directions: **1)** Take the square root of both sides of the equation.(on the right side, put ± in front of the number) **2)** Solve for the two values of x.

Examples

a. $(x - 4)^2 = 9$
$x - 4 = \pm 3$
$x - 4 = +3 \; ; \; x - 4 = -3$
$\underline{+4 \qquad +4 \qquad +4 \qquad +4}$
$x \quad = +7 \; ; \; x \quad = +1$

b. $(x + \frac{9}{2})^2 = \frac{169}{4}$
$x + \frac{9}{2} = \pm \frac{13}{2}$
$x + \frac{9}{2} = \frac{13}{2} \; ; \; x + \frac{9}{2} = -\frac{13}{2}$
$x = \frac{4}{2} \qquad ; \; x = -\frac{22}{2}$
$x = 2 \qquad ; \; x = -11$

Work the third step of completing the binomial square

1. $(x - 1)^2 = 16$

2. $(x + 5)^2 = 36$

3. $(x - 8)^2 = 25$

4. $(x + \frac{7}{2})^2 = \frac{25}{4}$

5. $(x + \frac{11}{2})^2 = \frac{49}{4}$

6. $(x - \frac{3}{2})^2 = \frac{1}{4}$

7. $(x + \frac{15}{2})^2 = \frac{81}{4}$

8. $(x - 3)^2 = 81$

9. $(x - 7)^2 = 100$

10. $(x + 1)^2 = 9$

11. $(x + \frac{13}{2})^2 = \frac{25}{4}$

12. $(x - 10)^2 = 4$

Completing the Binomial Square Continued - put all three steps together to solve the equation.

Example

$$x^2 - 12x + 35 = 0$$
$$\underline{ - 35 \quad -35}$$
$$x^2 - 12x + \underline{} = -35$$
$$x^2 - 12x + \underline{36} = -35 + 36$$
$$(x - 6)^2 = 1$$
$$x - 6 = \pm 1$$

$$x - 6 = +1 \quad ; \quad x - 6 = -1$$
$$\underline{+6 \quad +6} \quad ; \quad \underline{+6 \quad +6}$$
$$x = +7 \quad ; \quad x = +5$$

Solve by completing the binomial square

1. $x^2 + 8x - 20 = 0$

3. $x^2 - 12x - 13 = 0$

2. $x^2 - 11x + 30 = 0$

4. $x^2 + 3x + 2 = 0$

Solving Quadratic Equations by Completing the Binomial Square Continued

5. $x^2 + 8x - 9 = 0$

8. $x^2 + 7x + 12 = 0$

6. $x^2 - 9x + 20 = 0$

9. $x^2 - 5x + 6 = 0$

7. $x^2 - 10x + 21 = 0$

10. $x^2 + 4x - 21 = 0$

Writing Equations Given the Roots - a quadratic equation can be written if the roots are known. Follow these steps: **1)** Pair each root with **x** and put each inside parentheses, using the opposite sign. Example: $x = -5$, $(x + 5)$ **2)** Multiply the binomials using the FOIL method. **3)** Equate the trinomial to zero.

Examples

a. $x = 4$; $x = -7$
    $(x - 4)(x + 7) = 0$
    $x^2 + 7x - 4x - 28$
    $x^2 + 3x - 28 = 0$

b. $x = \frac{3}{4}$; $x = \frac{5}{2}$
    $(4x - 3)(2x - 5)$
    $4x^2 - 20x - 6x + 15$
    $4x^2 - 26x + 15 = 0$

[note: when going from the roots to the factors, the denominator moves to the left side of the binomial]

Write the equation given the roots

1. $x = 10$; $x = -1$

2. $x = -5$; $x = 4$

3. $x = \frac{5}{7}$; $x = -\frac{2}{3}$

4. $x = -\frac{3}{10}$; $x = \frac{5}{4}$

5. $x = 4$; $x = 9$

6. $x = -\frac{4}{5}$; $x = -\frac{6}{5}$

7. $x = -3$; $x = 8$

8. $x = 1$; $x = 5$

9. $x = \frac{5}{6}$; $x = -2$

10. $x = 9$; $x = \frac{1}{5}$

Chapter 14 Review

1. Solve for x:
   $7x^2 + 28x = 0$

2. Solve by factoring:
   $x^2 - 12x + 35 = 0$

3. Solve by using the quadratic formula:
   $x^2 - 6x - 27 = 0$

4. Write the equation given the roots:
   $x = -3 \; ; \; x = 11$

5. Solve by completing the (binomial)^2:
   $x^2 - 8x + 15 = 0$

6. Write the equation given the roots:
   $x = \frac{3}{2} \; ; \; x = -\frac{5}{6}$

7. Solve by factoring:
   $6x^2 + x - 2 = 0$

8. Solve for x:
   $(3x - 7)(x - 2) = 0$

Chapter 14 Review Continued

9. Solve by completing the ((binomial)^2:
   $x^2 + 6x - 16 = 0$

10. Solve by using the quadratic formula:
    $2x^2 + 13x + 6 = 0$

11. Solve for x:
    $5x(x + 2) = 0$

12. Write the equation given the roots:
    $x = 5$; $x = \frac{5}{3}$

13. Solve by factoring:
    $x^2 - 12x + 32 = 0$

14. Solve by completing the (binomial)^2:
    $x^2 + 7x - 8 = 0$

15. Solve for the variables:
    $(x)(y) = 0$

16. Solve by factoring:
    $x^2 - 14x = 15$

Cumulative Review 12

Simplify:

1. $\sqrt{81y^7}$

2. $3\sqrt{7}(4\sqrt{2} - \sqrt{6})$

3. $(\sqrt{5} - \sqrt{3})(\sqrt{5} + \sqrt{3})$

4. $\sqrt{\frac{5}{7}}$

5. Solve: $\sqrt{x+2} = 6$

6. Find the slope:
   $y = -4x + 7$

7. Write the equation given $(-4, 3)(-6, 9)$

8. Solve by addition:
   $2x - y = 2$
   $x + 2y = 11$

9. Solve: $\frac{3}{4} = \frac{x+1}{x-6}$

10. Multiply: $\frac{a-5}{a^2-2a-8} \times \frac{3a^2-10a-8}{3a+2}$

11. Solve by factoring: $x^2 + 11x = -18$

12. Factor: $3ay - 9y + 2a^2 - 6a$

13. Solve: $5x - 3x + x = 16 - 2 + 7$

14. Multiply: $3x^2 + 2x + 1$
    $\underline{3x - 2}$

15. Simplify: $6^2 - 50 \div 5^2 + 4 \bullet 5$

Chapter 15
Math Riddles

Introduction - Many math students dislike "word problems". This workbook will deal with "math riddles" instead of word problems. In a riddle there is always a question such as "What number am I thinking about?" or "How many coins do I have?"

Try to determine the answer to these math riddles.
What number am I thinking of?

1. It is 5 larger than 15.

2. It is 10 more than 25.

3. It is 8 smaller than 18.

4. It is 15 less than 40.

5. When I add 15 to my number, I get 35.

6. When I subtract 6 from my number, I get 30.

7. When I add 9 to my number, I get 40.

8. When I reduce my number by 20, I get 25.

9. When I multiply my number by 2, I get 14.

10. When I divide my number by 3, I get 10.

11. It is 5 less than 20.

12. It is 11 smaller than 30.

13. It is 8 more than 23.

14. It is 12 larger than 11.

15. When I add 15 to my number, I get 45.

16. When I reduce my number by 10, I get 28.

17. When I multiply my number by 5, I get 30.

18. When I subtract 20 from my number, I get 15.

19. When I divide my number by 9, I get 4.

20. It is 11 more than 7.

Phrases Used in Working Math Riddles - there are certain phrases that mean add, subtract and equal to:

<u>Add</u>	<u>Subtract</u>	<u>Equal to</u>
greater than	less than	is the same as
larger than	smaller than	gives you an answer of
more than	fewer than	is
exceeds a number by	diminished by	gives
added to	subtracted by	is equivalent to
a "plus" sign	subtracted from	to get
increased by	a "negative" sign	I get
raised by	decreased by	gives you the same
	lowered by	result as
	reduced by	is equal to

Five Steps in Solving Math Riddles - to solve math riddles, follow these steps:
    1. Read the riddle until you know what it is that the riddle expects you to find.
    2. Identify what you're trying to find and equate it to "x".
    3. Use "x" to identify anything else you might be looking for.
    4. Write an equation following the instructions of the riddle.
    5. Solve the equation. This gives you the answer to the riddle.

There are several forms of math riddles:
    1. One solution riddles   5. Quadratic equations
    2. Two or more solutions riddles   6. Rational equations
    3. One variable equations   7. Radical equations
    4. Two variable equations

Math Riddles With One Solution

Examples

a. Find the number I'm looking for. If I multiply my number by 3, I get 21.

x = number I'm looking for
$3x = 21$
$\dfrac{3x}{3} = \dfrac{21}{3}$
$x = 7$

b. Find my age. If I add 12 to my age, I get 20.

x = my age
$x + 12 = 20$
$-12\ \ -12$
$x = 8$

Solve:

1. Find the number I'm looking for. If I add 16 to my number, I get 35.

2. Find my age. If I add 47 to my age, I get 61.

3. Find my test score. If I multiply it by 4, I get 320.

4. Find my weight. If I multiply it by 7, I get 784.

5. Find the points scored. When I add 16 to my points, I get 33.

6. Find the money in my pocket. When I multiply my money by 6, I get $5.10.

7. Find the number of rainy days last month. If you add 11 to them, you get 18.

8. Find the number I'm looking for. If I multiply it by -4, I get -108.

9. Find the number of votes cast. When you add 265 to the votes, you get 943.

10. Find the number of freshmen. When I add 149 to the number, I get 387.

More One Solution Math Riddles

Examples

a. What is my age? If I subtract 15 from my age, I get 33.

x = my age
$x - 15 = 33$
$+ 15 + 15$
$x = 48$

b. What is my number? If I divide my mumber by 7, I get 6.

x = my number
$\frac{x}{7} = 6$
$\frac{7}{1} \cdot \frac{x}{7} = 6 \cdot 7$
$x = 42$

What is my age?

1. If I subtract 10 from my age, I get 25.

2. When I divide my age by 3, I get 7.

3. If you subtract 25 from my age, you get 61.

4. When you divide my age by 11, you get 6.

5. If I divide my age by 5, I get 10.

6. When you subtract 20 from my age, you get 43.

7. When I subtract 19 from my age, I get 6.

8. If you divide my age by -17, you get -2.

9. If you subtract 3 from my age to get 7.

10. When I divide my age by -6, I get -8.

11. If I divide my age by 10, I get 4.

More One Solution Math Riddles

Examples

a. What is my average?
   If I subtract my average from 98, I get 14.

   x = my average
   $98 - x = 14$
   $-98 \qquad -98$
   $-x = -84$
   $x = 84$

b. What is my average?
   If I multiply my average by 4, then add 7, I get 307.

   $4x + 7 = 307$
   $\quad -7 \quad -7$
   $4x = 300$
   $\overline{4} \qquad \overline{4}$
   $x = 75$

What is my weight?

1. If I subtract my weight from 200, I get 35.

2. If I multiply my weight by 3 and add 26, I get 431.

3. If I subtract my weight from 340, I get 218.

4. If I multiply my weight by 7 and add 73, I get 1,333.

5. If I subtract my weight from 176, I get 19.

6. If I multiply my weight by 11 and add 31, I get 2,374.

7. When I multiply my weight by -5 and add 16, I get -514.

8. When I subtract my weight from 143, I get -14.

9. When I subtract my weight from 219, I get 3.

10. When I multiply my weight and 7, then add 20, I get 1,217.

225

More One Solution Math Riddles

Examples

a. How many students made an A? When I multiply the number by 3 and subtract 8, I get 142.

x = number of A's
$3x - 8 = 142$
$+ 8 = + 8$
$3x = 150$
$\frac{3x}{3} = \frac{150}{3}$
$x = 50$

b. How many students made an A? If I multiply the number by 5, then multiply that product by 6, I get 300.

x = number of A's
$6(5x) = 300$
$30x = 300$
$\frac{30x}{30} = \frac{300}{30}$
$x = 10$

How many high schools have Science Clubs?

1. If I multiply the number by 5 and subtract 25, I get 30.

2. If I multiply the number by 3 and times the product by 6, I get 72.

3. If I multiply the number by 7 and subtract 45, I get 137.

4. If I multiply the schools by 4 then times the product by 3, I get 1200.

5. If I multiply the schools by 6 then subtract 104, I get 340.

6. If I multiply the schools by -5 and multiply the product by -2, I get 210.

7. If I multiply the number by -9 then times the product by -4, I get 180.

8. If I multiply the number by 10 and subtract 75, I get 385.

More One Solution Math Riddles

Examples

a. What is my number? The product of 3 and the sum of my number and 7 is equal to 30.

$$x = \text{my number}$$
$$3(x + 7) = 30$$
$$3x + 21 = 30$$
$$-21 \quad -21$$
$$\frac{3x}{3} = \frac{9}{3}$$
$$x = 3$$

b. What is my number? The product of 2 and the difference between my number and 12 is equal to (-60).

$$x = \text{my number}$$
$$2(x - 12) = -60$$
$$2x - 24 = -60$$
$$+24 \quad +24$$
$$\frac{2x}{2} = \frac{-36}{2}$$
$$x = -18$$

What is the number I'm thinking about?

1. The product of three and the sum of my number and 15 is equal to 108.

2. The product of 6 and the difference between my number and 8 equals 12.

3. The product of 12 and the sum of my number and 18 is equal to 336.

4. The product of 5 and the difference between my number and 35 equals 10.

5. The product of 4 and the difference between my number and 17 equals -8.

6. The product of 7 and the sum of my number and 104 equals 763.

7. Four times the sum of my number and 97 is equal to 488.

8. Six times the difference between my number and 5 equals 30.

More One Solution Math Riddles
Examples

a. What is my age? If I multiply my age by 3 and subtract 14, I get the same as when I add 20 to my age.

$$x = \text{my age}$$
$$3x - 14 = x + 20$$
$$\underline{-x + 14 \quad\quad -x + 14}$$
$$\frac{2x}{2} = \frac{+34}{2}$$
$$x \quad\quad 17$$

b. Find my age. If I multiply my age by 8 and subtract 4, I get the same as when I multiply my age by 4 and add 8.

$$x = \text{my age}$$
$$8x - 4 = 4x + 8$$
$$\underline{-4x + 4 \quad\quad -4x + 4}$$
$$\frac{4x}{4} = \frac{12}{4}$$
$$x = 3$$

How many points did Penn State score?

1. If I multiply the points by 6 and subtract 9, I get the same as when I add 46 to the points.

2. If I multiply the points by 10 and subtract 5, I get the same as when I multiply the points by 5 and add 10.

3. When I multiply the points by 5 and subtract 9, I get the same as when I add 11 to the points.

4. When I multiply the points by 4 and subtract 20, I get the same as when I multiply the points by 2 and add 6.

5. If I times the points by 7 and reduce them by 11, I get the same as when I times the points by 3 and increase them by 17.

6. When I times the points by 3 and take away 10, I get the same as when I increase the points by 14.

7. When I multiply the points by 6 and add 7, I get the same as when I add 37 to the number of points.

8. If I times the points by 6 and increase them by 5, I get twice the points increased by 77.

More One Solution Math Riddles
Examples
What percent of boys like nacho flavored ice cream?

a. If I subtract 4 from the percent and multiply the difference by 3, I get the same as when I subtract 2 from the percent.
   x = the percent of boys
   $3(x - 4) = x - 2$
   $3x - 12 = x - 2$
   $-x + 12 \quad -x + 12$
   $\dfrac{2x}{2} = \dfrac{+10}{2}$
   $x = 5$

b. The sum of the percent divided by 3 and the percent divided by 6 equals 1.
   x = the percent of boys
   $\dfrac{x}{3} + \dfrac{x}{6} = 1$

   $\dfrac{6(x) + 3(x)}{18} = 18(1)$
   $6x + 3x = 18$
   $\dfrac{9x}{9} = \dfrac{18}{9}$
   $x = 2$

How many soccer players take a math course?

1. If I subtract 10 from the number and multiply the difference by 4, I get the same as when I subtract 7 from the number.

2. The sum of the number divided by 3 and the number divided by 5 equals 8.

3. When I add 3 to the number and multiply the sum by 2, I get the same as when I add 36 to the number.

4. The sum of my number divided by 2 and my number divided by 3 is 10.

5. When I decrease my number by 6 and multiply the sum by 4, I get the same as when I subtract 9 from my number.

6. The sum of the number divided by 4 and the number divided by 2 is equal to 12.

7. If I decrease the number by 5 and multiply the difference by 6, I get the same as decreasing the number by 10.

8. The sum of the number divided by 3 and the number divided by 5 is 12.

More One Solution Math Riddles

Examples

a. How many games did Florida win? If I add 25 to that number, then reduce the number by 16, I get 32.

x = games won
$x + 25 - 16 = 32$
$x + 9 = 32$
$-9 -9$
$x = 23$

b. How many games did Alabama win? The number equals its opposite added to 36.

x = games won
$x = -x + 36$
$+x +x$
$2x = 36$
$2 2$
$x = 18$

Find the temperature outside today.

1. If I add 45 degrees to the temperature, then reduce the number by 16, I get 16 degrees.

2. The temperature equals its opposite added to 60 degrees.

3. If I add 31 degrees to the temperature, then reduce it by 14, I get 45 degrees.

4. The temperature equals its opposite added to 46 degrees.

5. If I increase the temperature by 60 degrees, then decrease it by 29 degrees, I get 51 degrees.

6. The temperature equals its opposite added to 144 degrees.

7. If you add 19 degrees to the temperature, and subtract 14 degrees, you get 75 degrees.

8. If you increase the temperature by 35 degrees, and reduce it by 48 degrees, it will be 65 degrees.

More One Solution Math Riddles

Examples

a. Find my bank balance. My bank balance equals its opposite decreased by $40.

$$x = \text{my bank balance}$$
$$x = -x - 40$$
$$+x \quad +x$$
$$\frac{2x}{2} = \frac{-40}{2}$$
$$x = -20$$

b. Find my bank balance. Two times my bank balance equals my bank balance added to $16.

$$x = \text{my bank balance}$$
$$2x = x + 16$$
$$-x \quad -x$$
$$x = +16$$

Find my bank balance

1. My bank balance equals its opposite decreased by $200.

2. Three times my balance equals my balance added to $100.

3. My balance equals its opposite decreased by $16.

4. Five times my balance equals my balance added to $360.

5. My balance equals its opposite reduced by $96.

6. My balance equals its opposite subtracted by (-$40).

7. Eight times my bank balance is the same as my balance added to $700.

8. Four times my balance equals my balance raised by $3000.

More One Solution Math Riddles.

Examples

a. What is my age? Six times my age is 40 more than two times my age.

$$x = \text{my age}$$
$$6x = 2x + 40$$
$$-2x = -2x$$
$$\frac{4x}{4} = \frac{40}{4}$$
$$x = 10$$

b. What is my age? If I add 30 to one fourth my age, I get 50.

$$x = \text{my age}$$
$$\tfrac{1}{4}x + 30 = 50$$
$$-30 \quad -30$$
$$4 \cdot \tfrac{1}{4}x = 20 \cdot 4$$
$$x = 80$$

Find the average temperature in March.

1. Eight times the temperature is 90 more than 5 times the temperature.

2. If I add 25 degrees to $\tfrac{1}{3}$ the temperature, I get 40 degrees.

3. Six times the temperature is 64 degrees more than 2 times the temperature.

4. Seven times the temperature is 40 degrees more that 3 times the temperature.

5. The temperature multiplied by 9 equals 63 degrees added to the temperature multiplied by 2.

6. If I add 13 to $\tfrac{1}{4}$ the temperature I get 35 degrees.

7. Twenty more than $\tfrac{1}{2}$ the temperature equals 41 degrees.

8. Thirty-two more than one fifth the temperature is 45 degrees.

More One Solution Math Riddles

Examples

a. Find the number. The square root of the number is equal to 8.

$$x = \text{the number}$$
$$\sqrt{x} = 8$$
$$(\sqrt{x})^2 = (8)^2$$
$$x = 64$$

b. Find the number. The square root of the number that is reduced by 3 is the same as 7.

$$x = \text{the number.}$$
$$\sqrt{x-3} = 7$$
$$(\sqrt{x-3})^2 = (7)^2$$
$$x - 3 = 49$$
$$+3 \quad +3$$
$$x = 52$$

Find my number.

1. The square root of my number is equal to 10.

2. The square root of my number that has been reduced by 5 is 9.

3. The square root of my number is the same as 4.

4. The square root of my number that has been decreased by 3 is equal to 8.

5. Fifteen is the same as the square root of my number.

6. The square root of my number that has been increased by 6 is equal to 10.

7. Twelve is the same as the square root of my number.

8. The square root of my number that has been increased by 2 is 7.

Review of One Solution Math Riddles

What number am I thinking of?

1. If I multiply my number by 7 I get 56.

2. If I add 25 to my number I get -6.

3. If I subtract 72 from my number, I get -10.

4. If I divide my number by 11, I get 8.

5. If I subtract my number from -35, I get 16.

6. If I multiply my number by 6 and add -10 to get 20.

7. If I times my number by 9 and lower it by 10 to get -100.

8. If I multiply my number by 4 and times the product by 8 to get 96.

9. If I increase my number by 17 and double the sum, I get 84.

10. If I decrease my number by 19 and triple the difference to get 123.

11. If I raise my number by 30 and lower it by 45 to get -23,

How many coins do I have?

12. Ten times my number of coins is 64 more than 2 times my number.

Review of One Solution Math Riddles Continued

13. Five is the same as the square root of my number of coins.

14. The number of my coins equals it opposite decreased by -30

15. When I add 5 to the number of coins and multiply the sum by (-6), I get the same as when I take 51 from that number.

16. The sum of my coins divided by 3 and my coins divided by 7 is equal to 10.

17. Thirty-one more than $\frac{1}{6}$ the number of coins is equal to 37.

18. The square root of my number reduced by 11 is the same as 3.

19. Doubling my number of coins is equal to adding 44 to the number of coins.

20. When I multiply the number of coins by 9 and add 10, I get the same as when I multiply the number by 10 and subtract 9.

Math Riddles With Two Solutions - when solving math riddles with more than one solution, x must be identified and the other solution(s) must also be identified. Notice that x must be used in identifying the other solution(s). As illustrated in the example, usually there is more than one way to solve two solution math riddles.

Example
One number is 6 larger than another and their sum is 54. Find both numbers.

a.
$$x = \text{the smaller number}$$
$$x + 6 = \text{the larger number}$$
$$x + (x + 6) = 54$$
$$2x + 6 = 54$$
$$-6 = -6$$
$$\frac{2x}{2} = \frac{48}{2}$$
$$x = 24 \text{ (the smaller \#)}$$
The larger # is x + 6, (24 + 6), 30

b.
$$x = \text{the larger number}$$
$$x - 6 = \text{the smaller number}$$
$$x + (x - 6) = 54$$
$$2x - 6 = 54$$
$$+6 \quad +6$$
$$\frac{2x}{2} = \frac{60}{2}$$
$$x = 30 \text{ (the larger \#)}$$
The smaller # is x - 6, (30 - 6) 24

Note that in both problems the smaller # is 24 and the larger is 30

Find my numbers. (Work each problem 2 ways.)

1. One number is 16 larger than another and their sum is 48.

2. One number is 35 greater than a second and they add up to 99.

3. A number is 30 less than a second and they add up to 84.

236

Work these problems only one way.

Example
Joe scored 16 points more than Bob. Together, they scored 40 points. How many points did each score?

$$x = \text{Bob's points}$$
$$x + 16 = \text{Joe's points}$$
$$x + (x + 16) = 40$$
$$2x + 16 = 40$$
$$-16 \quad -16$$
$$\frac{2x}{2} = \frac{24}{2}$$
$$x = 12 \quad (\text{Bob's points})$$

Joe scored x+16, (12 + 16), 28 points

---

1. Shirley made 10 points higher on a test than Paul did. Their grades add up to 180 points. Find each test score.

2. Linda played 2 hours longer than she studied. She spent 10 hours doing both. How long did she spend doing each?

3. Jody s 15 years older than Jane. Their ages add up to 57. Find their ages.

4. Glen weighs 38 pounds less than Murray. Together they weigh 392 pounds. Find the weight of each man.

5. Dodie is 4 inches shorter than Al. Their heights add up to 146 inches. Find the height of each.

More Two Solution Math Riddles

Examples

a. Find the numbers. One number is 12 times another and their difference is 44.

x = the smaller number
$12x$ = the larger
$12x - x = 44$
$\underline{11x} = \underline{44}$
$11 \quad 11$
$x = 4$ (the smaller #)
The larger # is 12x, (12•4), 48.

b. Find the numbers. The difference between the 2 numbers is 40. If the larger is subtracted from twice the smaller, the difference is 63.

x = the smaller number
$x + 40$ = the larger number
$2x - (x + 40) = 63$
$2x - x - 40 = 63$
$x - 40 = 63$
$\underline{+40 \quad +40}$
$x = 103$ (smaller #)
The larger # is x + 40, (103 + 40), 143

Find the numbers.

1. One number is 10 times another and their difference is 63.

2. The difference between two numbers is 30. If the larger is subtracted from twice the smaller, the difference is 6.

3. One number is 6 times another and their difference is 40.

4. The difference between 2 numbers is 8. If the larger is subtracted from three times the smaller the difference is 100.

5. One number is 4 times another and their difference is -27.

6. The difference between 2 numbers is 72. When the greater is subtracted from 5 times the lesser, the difference is 36.

More Two Solution Math Riddles - consecutive integer problems
1. Consecutive integers are integers that follow one another: ..., -3, -2, -1, 0, 1, 2, 3,....
2. Consecutive <u>even</u> integers are: ...,-2, 0, 2, 4, 6,... (each is 2 away from the next)
3. Consecutive <u>odd</u> integers are: ...,-3, -1, 1, 3, 5,... (each is also 2 away from the next)

Example

Name three consecutive <u>odd</u> integers, whose sum of the first and third is 78.

$$x = \text{the first integer}$$
$$x + 2 = \text{the second integer}$$
$$x + 4 = \text{the third integer}$$
$$x + (x + 4) = 78$$
$$2x + 4 = 78$$
$$-4 \quad -4$$
$$\frac{2x}{2} = \frac{74}{2}$$
$$x = 37 \text{ (the first integer)}$$

The third integer is x + 4, (37 + 4), 41

---

1. The sum of two consecutive integers is 105.

2. Name three consecutive odd integers, whose sum of the 1st and 3rd is 94.

3. The sum of two consecutive integers is 175. Find the integers.

---

4. The sum of three consecutive integers is -54. Find them.

5. Name three consecutive odd integers whose sum of the 2nd and the 3rd is 68.

6. Name three consecutive even integers whose sum of the 1st and 3rd is 84.

7. The sum of three consecutive integers is 24. Find the integers.

8. The sum of two consecutive integers is 191. Find them.

239

More Two Solution Math Riddles

Example
At a picnic, there were 4 more girls than twice the number of boys. Together, there were 124 students. Find the number of boys and girls.

$$
\begin{aligned}
x &= \text{number of boys} \\
2x + 4 &= \text{number of girls} \\
x + (2x + 4) &= 124 \\
3x + 4 &= 124 \\
-4 \quad &\quad -4 \\
\frac{3x}{3} &= \frac{120}{3} \\
x &= 40 \quad (\text{\# of boys})
\end{aligned}
$$

The number of girls is 2x + 4, (2•40 + 4), 84

-----

1. Kathie is 2 years older than three times Ashley's age. Together their ages add up to 98. Find Kathie and Ashley's age.

2. There are 5 more right handed students than 4 times the number of left handed students. There are 75 students. How many are left handed?

3. Jacob earned $10 more than twice what Mort earned last week. Together they earned $61. How much did each earn?

4. In an election, the loser got 3 fewer votes than 5 times the winner's votes. Together they got 1,113 votes. How many did each get?

5. Jill got 8 fewer problems correct than 3 times Lisa's correct problems. Together they got 72 problems correct. How many did each get correct?

More Two Solution Math Riddles - this particular problem can be worked 3 different ways. Note, the third way is to use two variables and solve as simultaneous equations.

Example

The sum of the two numbers is 16. One number is 3 times the other.

x = smaller number	x = the smaller number	x = smaller number
$3x$ = the larger	$16 - x$ = the larger number	y = larger number
$x + 3x = 16$	$3x = 16 - x$	$3x = y$
$4x = 16$	$+x \quad +x$	$3x - y = 0$
$\dfrac{4x}{4} = \dfrac{16}{4}$	$\dfrac{4x}{4} = \dfrac{16}{4}$	$x + y = 16$
$x = 4$	$x = 4$	$4x = 16$
		$x = 4$

In all three methods the smaller number is 4 and the larger number is 12

Solve the riddles using any method

1. The sum of two numbers is 48. One number is 5 times the other. Find them.

2. The sum of two numbers is 30. One number is twice the other. Find them.

3. The total of two numbers is 25. One number is 4 times the other. Find them.

4. The total of two numbers is 80. One number is 7 times the other. Find them.

5. Two numbers add up to 36. One is 8 times the other. Find them.

6. Two numbers total 77. One number is 6 times the other. Find them.

More Two Solution Math Riddles - the first example becomes a quadratic equation that is solved by factoring and the second example sets up as two linear equations that are solved by elimination. (addition)

Examples

a. Name 2 consecutive odd integers whose product is 99.

x = the smaller integer
$x + 2$ = the larger integer
$x(x + 2) = 99$
$x^2 + 2x = 99$
$\underline{ - 99 -99}$
$x^2 + 2x - 99 = 0$ (factor)
$(x + 11)(x - 9) = 0$
$x = -11, \; x = 9$
smaller integer = 9, larger = 11
　　　　or
smaller integer = -11, larger = -9

b. The sum of two integers is 48 and their difference is 10.

x = the larger integer
y = the smaller integer
$x + y = 48$
$\underline{x - y = 10}$
$2x = 58$
$\overline{22}$
$x = 29$

The larger integer, x is 29
The smaller, y (29 + y = 48) is 19

Set Up and Solve as a Quadratic Equation or as Simultaneous Equations

1. Name 2 consecutive <u>even</u> integers whose product is 48.

2. Name 2 consecutive <u>odd</u> integers whose product is 35.

3. Name 2 consecutive integers whose product is 42.

4. The sum of two numbers is 60 and their difference is 16. Find them.

Two Solution Math Riddles Continued

5. The sum of two numbers is 45 and their difference is 31. Find them.

6. The sum of two numbers is 116 and their difference is 46. Find them.

7. Name 2 consecutive odd integers whose product is 63.

8. The sum of two numbers is 76 and their difference is 12. Find them.

9. Name two consecutive odd integers whose product is 15.

10. Two numbers add to give you 50 and subtract to give you 8. Find them.

11. The sum of two numbers is 64 and their difference is 30. Find them.

12. Name two consecutive integers whose product is 56.

Review of Two Solution Math Riddles

1. One number is 18 less than another and their sum is 46. Find them.

2. One number is 29 larger than another and they add up to 107. Find them.

3. Billy made 17 points more on a test than Philip did. Their test grades add up to 175. What was each boy's grade?

4. Linda is 6 inches shorter than Chris. The sum of their heights is 132 inches. Find the height of each.

5. One number is 7 times another and their difference is 60. Find the numbers.

6. Two numbers differ by 18. If the larger is subtracted from 4 times the smaller, the result is 21. Find the numbers.

7. The sum of two consecutive integers is 217. Find the integers.

8. Name three consecutive even integers whose sum is 156.

9. Whit studied 30 minutes more than 3 times what Nick studied. Together they studied a total of 210 minutes. How long did each study?

10. The sum of two numbers is 48. One number is twice the other. Find the numbers.

Review of Two Solution Math Riddles Continued

11. Name 2 consecutive even integers whose product is 80.

12. The sum of two numbers is (-19) and their difference is 35. Find them.

13. If I add 27 pounds to my weight, I weigh 203.

14. If I divide my math average by 7, I get 12.

15. When I multiply my age by 10 and add 46, I get 536.

16. If I multiply my number by 7, then times the product by 4, I get 84.

17. The product of 5 and the sum of my number and 14 is 110.

18. My number equals its opposite added to 52.

19. Three times my number of pets equals my number of pets added to 10.

20. Six times my number of Algebra students reduced by 20 is the same as 4 times my number of Algebra students added to 22.

21. If I add 21 to 1/3 my weight I get 72.

More Two Solution Math Riddles

Examples

a. Jason is 8 years older that Randy. In 5 years the sum of their ages will be 43. How old is each boy?

x = Randy's age
$x + 8$ = Jason's age
$x + 5$ = Randy's age in 5 years
$x + 13$ = Jason's age in 5 years

$$(x + 5) + (x + 13) = 43$$
$$2x + 18 = 43$$
$$-18 \quad -18$$
$$\frac{2x}{2} = \frac{25}{2}$$
$$x = 12\tfrac{1}{2} \text{ Randy's age}$$

Jason's age is $(12\tfrac{1}{2} + 8)$, $20\tfrac{1}{2}$

b. Nelson weighs 20 pounds more than Jay. Tony weighs 130 pounds less than twice as much as Jay. If Nelson and Tony weigh the same, how much does each boy weigh?

x = Jay's weight
$x + 20$ = Nelson's weight
$2x - 130$ = Tony's weight

$$2x - 130 = x + 20$$
$$-x + 130 \quad -x + 130$$
$$x = 150 \text{ Jay's weight}$$

Nelson's weight is $(150 + 20)$, 170
Tony's weight is $(2 \bullet 150 - 130)$, 170

Solve

1. Jody is 5 months older than Bo. In 8 months, the sum of their months will be 47. Find how many months old each is.

2. Bobby weighs 30 pounds more than Matt. "T" weighs 210 pounds less than three times what Matt weighs. Bobby and "T" weigh the same. Find the weight of each.

Two Solution Math Riddles Continued

3. Bill scored 15 more points than Shag. Skinny scored 40 points less than twice Shag's points. Bill and Skinny scored the same number of points. Find the points each scored.

4. Rick is 10 years older than Randy. In 12 years the sum of their ages will be 60. Find their ages.

5. Emily is 3 years older than Lynn. In 7 years the sum of their ages will be 37. Find their ages.

6. Tim picked 12 more bushels of corn than Seth. Brent picked 160 fewer than 3 times what Seth picked. Brent and Tim picked the same number of bushels. How many did each boy pick?

More Two Solution Math Riddles

Examples

a. I have 4 more dimes than nickels and 1 less quarter than nickels. I have 18 coins. How many of each coin do I have, and how much money do I have?

$$x = \text{number of nickels}$$
$$x + 4 = \text{number of dimes}$$
$$x - 1 = \text{number of quarters}$$

$$x + (x + 4) + (x - 1) = 18$$
$$3x + 3 = 18$$
$$-3 \quad -3$$
$$\frac{3x}{3} = \frac{15}{3}$$
$$x = 5 \text{ Nickels}$$

The number of dimes is (5 + 4), 9
The number of quarters is (5 - 1), 4

I have $5(.05) + 9(.10) + 4(.25) = \$2.15$

b. I have 2 more dimes than nickels, and 3 fewer quarters than nickels. If I have $1.85, how many of each coin do I have?

$$x = \text{number of nickels}$$
$$x + 2 = \text{number of dimes}$$
$$x - 3 = \text{number of quarters}$$
$$.05x = \text{the value of the nickels}$$
$$.10(x + 2) = \text{the value of the dimes}$$
$$.25(x - 3) = \text{the value of the quarters}$$

$$.05(x) + .10(x + 2) + .25(x - 3) = \$1.85$$
$$5x + 10x + 20 + 25x - 75 = 185 \text{ (mult. by 100)}$$
$$40x - 55 = 185$$
$$+55 \quad +55$$
$$\frac{40x}{40} = \frac{240}{40}$$
$$x = 6 \text{ number of nickles}$$

The number of dimes is (6 + 2), 8
The number of quarters is (6 - 3), 3

Solve

1. I have 6 more quarters than dimes and 2 fewer nickels than dimes. I have 25 coins. How many of each coin do I have and how much money do I have?

2. I have 4 more nickels than quarters and 3 fewer dimes than quarters. I have $1.90. How many of each coin do I have?

Two Solution Math Riddles Continued

3. Lesa has 5 more quarters than dimes, and 7 fewer nickels than dimes. She has 25 coins. How many of each coin does she have and how much money does she have?

4. Jack Thompson has 7 more quarters than dimes and 1 less nickel than dimes. He has $3.70. How many coins of each does he have?

5. Tasha has 3 more nickels than quarters and 2 fewer dimes than quarters. She has 13 coins. How much money does she have? (How many coins of each does she have?)

6. Juno has 4 fewer quarters than nickels and 9 more dimes than nickels. She has $3.10. How many coins of each does she have?

More Two Solution Math Riddles

Examples

a. Two planes leave Atlanta going in opposite directions. The first travels 500 mph and the second travels 350 mph. In how many hours will they be 3,400 miles apart?

t = the time in hours

Use the formula: rate • time = distance
Make a chart, using: r • t = d

	r	•	t	=	d
First Plane	500		t		500t
Second Plane	350		t		350t
total distance				=	850t
total distance				=	3,400 miles

$$\frac{850t}{850} = \frac{3,400}{850}$$

t = 4 hours

b. John ran 8 mph in a race. Gretchen, running 5 mph, took 1 hour longer to finish. How long did it take John to finish the race, and how long was the race?

t = the time it took John to finish
t + 1 = the time it took Gretchen

Using the formula: r•t = d, make a chart:

	r	•	t	=	d
John	8		t	=	8t
Gretchen	5		t+1	=	5(t+1)

John's distance = 8t
Gretchen's distance = 5(t + 1)
They ran the same race, therefore:

5(t + 1) = 8t
5t + 5 = 8t
-5t   - 5t

$$\frac{5}{3} = \frac{3t}{3}$$

$1\frac{2}{3}$ = t (in hours)

distance: 8t, $(8 \cdot 1\frac{2}{3}) = 13\frac{1}{3}$ miles

Solve

1. Two bikers left Palm Beach going in opposite directions. The first biker goes 35 mph and the second, 25 mph. In how many hours will they be 180 miles apart?

2. Clem biked 18 mph in a race. Annie, biking 15 mph took 4 hours longer. How long did it take Clem to finish the race and how long was the race?

Two Solution Math Riddles Continued

3. Two planes leave New York going in opposite directions. The first plane travels 600 mph and the second, 450 mph. In how many hours will they be 2,625 miles apart?

4. Jeff ran 5 mph in a race. Bert, running 3 mph, took 2 hours longer to finish. How long did it take Jeff to finish and how long was the race?

5. Two cars leave Orlando going in opposite directions. The first car goes 40 mph and the second car goes 60 mph. In how many hours will they be 500 miles apart?

6. Joe ran 9 mph in a race. Jack, running 6 mph, took 2 hours longer. How long did it take Joe to finish and how long was the race?

More Two Solution Math Riddles

Examples

a. Gomer invested three times as much money at 9% interest per year as he invested at 6%. How much did he invest at each rate if his interest income for the year was $528?

$\quad$ x = money invested at 6%
$\quad$ 3x = money invested at 9%
$\quad$ .06(x) = interest earned at 6%
$\quad$ .09(3x) = interest earned at 9%

.06(x) + .09(3x) = 528
(multiply by 100 to eliminate the decimals)
$\quad$ 6(x) + 9(3x) = 52800
$\quad$ 6x + 27x = 52800
$\quad$ $\frac{33x}{33}$ = $\frac{52800}{33}$
$\quad\quad$ x = $1600   invested at 6%
$\quad\quad$ 3x = $4800   invested at 9%

b. A Total of $8000 is invested, some at 5% and the rest at 9%. If the interest earned per year is $480 from both investments, how much was invested at each rate?

$\quad$ x = money invested at 5%
$\quad$ 8000 - x = money invested at 9%
$\quad$ .05(x) = interest earned at 5%
$\quad$ .09(8000 - x) = interest earned at 9%

.05(x) + .09(8000 - x) = 480
(multiply by 100 to eliminate the decimals)
$\quad$ 5(x) + 9(8000 - x) = 48000
$\quad$ 5x + 72000 - 9x = 48000
$\quad$ 72000 - 4x = 48000
$\quad$ - 72000   - 72000
$\quad\quad$ $\frac{-4x}{-4}$ = $\frac{-24000}{-4}$
$\quad\quad$ x = 6000   invested at 5%
$\quad\quad$ 8000 - x = 2000   invested at 9%

Solve

1. Jinx invested twice as much at 10% per year as she did at 8%. How much did she invest at each rate if her interest income for the year was $140?

2. A total of $4,000 is invested, some at 6% and some at 8%. If the interest earned per year is $300, how much was invested at each rate?

Two Solution Math Riddles Continued

3. Fat Pat invested twice as much at 8% per year than he did at 5%. How much did he invest at each rate if his interest income for the year was $630?

4. Shirley invested 3 times as much at 12% per year as she invested at 9%. If she earned $900 per year in interest income, how much did she invest at each rate?

5. Slim Jim invested $9,000, some at 8% and the rest at 10%. If the interest earned is $760, how much did he invest at each rate?

6. J.R. invested $7,000, some at 9% and the rest at 7%. If he earns $600 per year in interest, how much did he invest at each rate?

More Two Solution Math Riddles
Examples

a. How many pints of water must be added to 40 pints of 80% antifreeze solution to get a solution that is 70% antifreeze?

x = pints of water added
$x + 40$ = pints of water in new solution
80% = amount of antifreeze in old solution
70% = amount of antifreeze in new solution

$.8(40) = .7(x + 40)$
(multiply by 10 to eliminate the decimal)
$8(40) = 7(x + 40)$
$320 = 7x + 280$
$-280 \quad\quad -280$
$\dfrac{40}{7} = \dfrac{7x}{7}$
$5\tfrac{5}{7}$ = x (pints of water added)

b. How many ounces of a float that is 80% Dr. Pepper must be added to 24 ounces of a float that is 50% Dr. Pepper to get a float that is 60% Dr. Pepper?

x = ounces of 80% D.P. float to be added
$x + 24$ = total ounces in the new float
$.8(x)$ = D.P. in the solution that is added
$.5(24)$ = D.P. in the old float
$.6(x + 24)$ = D.P. in the new float

$.8(x) + .5(24) = .6(x + 24)$
(multiply by 10 to eliminate the decimal)
$8x + 5(24) = 6(x + 24)$
$8x + 120 = 6x + 144$
$-6x - 120 \quad -6x - 120$
$\dfrac{2x}{2} = \dfrac{24}{2}$
$x = 12$

Solve

1. How many pints of water must be added to 25 pints of 75% antifreeze solution to get a solution that is 60% antifreeze?

2. How many ounces of a float that is 75% Dr. Pepper must be added to 16 ounces of a float that is 60% Dr. Pepper to get a float that is 65% Dr. Pepper?

Two Solution Math Riddles Continued

3. How many quarts of water must be added to 16 quarts of 70% antifreeze solution to get a solution that is 50% antifreeze?

4. How many ounces of a float that is 60% root beer must be added to 12 ounces of a float that is 40% root beer to get a float that is 50% root beer?

5. How many ounces of 7 Up must be added to 12 ounces of 60% orange juice to get a solution that is 40% orange juice?

6. How many pints of 90% antifreeze must be added to 20 pints of 50% antifreeze to get a solution that is 70% antifreeze?

Chapter 15 Review

Find my number:

1. If I subtract my number from 45, I get (-13).

2. If I multiply my number by 7 and multiply the product by 3, I get 105.

3. If I decrease my number by 7 and double the difference, I get 48.

4. If I raise my number by 43 and lower it by 71, I get (-5).

5. Two runners leave Jonesboro going in opposite directions. The first is running 5 mph and the second is running 8 mph. In how many hours will they be 26 miles apart?

6. Name three consecutive even integers whose sum of the first and third is 84.

7. The difference beween two numbers is 17. If the larger is subtracted from 3 times the smaller, the difference is 63. Find the numbers.

8. Name two consecutive integers whose product is 72.

9. The sum of two integers is 76 and their difference if 14. Find the integers.

10. How many pounds of M & M's must be added to 10 pounds of 70% Nut Mix to get a mixture than is 60% Nut Mix.

11. Dr.Holcombe has 6 more dimes than quarters and 7 fewer nickels than quarters. If he has $6.65, how many of each coin does he have?

12. Bertha is 7 years older than Clem and the sum of their ages is 29. How old is each?

13. The sum of two numbers is 42. One number is 5 times the other. Find the numbers.

14. Annie invested twice as much at 5% per year as she did at 8%. How much did she invest at each rate if her interest income for the year was $360?

15. In an election the winner got 91 fewer votes than 3 times the loser. Together, they got 261 votes. How many votes did the loser get?

16. Tasha is 10 years older than Sammy. In 7 years, the sum of their ages will be 27. How old is each now?

Cumulative Review 13

1. Solve by using the Quadratic Formula:
$2x^2 + 3x - 5 = 0$
$x = \dfrac{-b \pm \sqrt{b^2 - 4ac}}{2a}$

2. Solve by completing the binomial square:
$x^2 - 4x - 21 = 0$

3. Simplify: $\sqrt{\dfrac{36}{49}}$

4. Simplify: $\dfrac{6\sqrt{50}}{2\sqrt{2}}$

5. Simplify: $4\sqrt{27} - 3\sqrt{75}$

6. Find the slope using $m = -\dfrac{A}{B}$
$4x - 7y = 11$

7. Write the equation given slope = 6 and y-intercept = -1

8. Solve by substitution: $3x - y = 2$
$x + y = 2$

9. Simplify: $\dfrac{x-4}{2} + \dfrac{2x+3}{3}$

10. Divide: $\dfrac{6a^2 b}{9c^2} \div \dfrac{10b^3 c}{4ac^2}$

11. Solve for x: $(2x + 1)(3x - 4) = 0$

12. Factor: $9x^2 - 49$

13. $2x + 1 \,\overline{\smash{\big)}\, 6x^2 + x - 1}$

14. Solve: $4(x - 3) = 11 - (x - 42)$

15. Subtract: $6x^2 - 5x + 13$
    $\underline{2x^2 - 3x - 11}$

16. $(2)(-3)(5) - (4)(-3)(-\frac{1}{2}) + (-7)(9)$

17. By substitution, determine if "7" is the solution to the equation: $5x - 11 = 2x + 9$

18. Simplify:
    $8x^2 - 2y^2 + xy - x^2 + 10xy - 3xy - 4x^2 + 6y^2$

19. Multiply: $(x + 9)(7x + 8)$

20. Multiply: $(3x^3y^5)^3(x^4y^3z^2)^4$

21. Divide: $\dfrac{-24x^6}{-8x^2}$

22. Simplify: $\dfrac{3^{-2}x^4y^{-1}}{4x^{-3}y^4}$

23. Solve: $\dfrac{5x}{3} + 7 = 22$

24. Put on the number line and give the solution set: $4 \leq x > -2$

    ─────────|─────────

    x = { 　　　　　　　　　　　　　　　　　　　　　　　　　}

Factor:

25. $x^2 - 11x + 18$

26. $5a^2 - 3ab$

27. $-2x^2 + 50$

28. Solve for x: $4x^2 + 28x = 0$

29. $\dfrac{x^2+7x-8}{x-1} \div (x+8)$

continue on the next page

30. Simplify: $\dfrac{5x-1}{4} - \dfrac{2x+7}{4}$

31. Solve for x: $\dfrac{3}{x-1} = \dfrac{2}{x+5}$

32. Solve for x and y by addition:
$3x + 5y = -4$
$4x - 3y = 14$

33. Put the linear equation in "slope-intercept" form:
$6x + 2y = 10$

34. Find the slope of the equation:
$y = -\dfrac{3}{8}x + \dfrac{2}{3}$

35. Write the equation given: (-1,6) (1,2)
(hint: you must find the slope and the y-intercept)

36. Solve for x: $\sqrt{3x-2} = 5$

37. Multiply: $(2\sqrt{5} + 3)(4\sqrt{5} - 1)$

38. Solve by graphing:
$x + y = 4$
$x - y = -2$

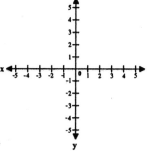

39. Solve by factoring: $5x^2 + 12x + 4 = 0$

40. Write the quadratic equation given the following roots: $x = -7, \ x = 5$